SAT* II:

MATHEMATICS
LEVELS IC, IIC

1998

BY ROBERT STANTON

SIMON & SCHUSTER

*SAT is a registered trademark of the College Entrance Examination Board, which is not affiliated with this book.

Kaplan Books
Published by
Kaplan Educational Centers and Simon & Schuster
1230 Avenue of the Americas
New York, New York 10020

Manufactured in the United States of America.
Published simultaneously in Canada.

Special thanks to: Jay Johnson, Evelyn Lontok, Krista Pfeiffer, and Robert Reiss.

Project Editor: Donna Ratajczak
Executive Editor: Del Franz
Managing Editor: Kiernan McGuire
Production Coordinator: Gerard Capistrano
Production Editor: Maude Spekes

August 1997
9 8 7 6 5 4 3 2 1

ISBN 0-684-84163-0
ISSN 1087-7924

CONTENTS

Introduction

How you use this book depends on which SAT II: Mathematics Subject Test you're planning to take and how much time you have. This next section will show you how to get the most out of your study time.

How to Use This Book

For more than fifty years, Kaplan has prepared students to take SATs. Our team of researchers and editors knows more about SAT preparation than anyone else, and you'll find their accumulated experience and knowledge in this book. As you work your way through the chapters, we'll show you precisely what knowledge and skills you'll need in order to do your very best on the SAT II: Mathematics Subject Tests. You'll discover the most effective way to tackle each type of question, and you'll reinforce your studies with lots of practice questions. At the back of the book you'll find four full-length, in-format tests with answer keys, scoring instructions, and detailed explanations. In addition, the Kaplan Advantage™ Stress Management System section contains helpful tips on beating test stress while you're preparing for the test and pulling off a victory on Test Day.

KAPLAN

The Best Prep

Kaplan's four full-length practice tests give you a great prep experience for the SAT II: Mathematics Subject Tests.

Get Ready to Prep

If possible, work your way through this book bit by bit over the course of a few weeks. Cramming the week before the test is not a good idea. You probably won't absorb much information if you try to pack it in at the last minute.

Start your preparation for the SAT II: Mathematics Subject Test by reading the Kaplan Advantage Stress Management System. The stress-relief tips will help you stay calm and centered. Keep using the tips that work for you before, during, and after the test.

Learn the Basics

The first thing you need to do is find out what's on the SAT II: Mathematics Tests. In the first section of this book, "The Basics," we'll provide you with background information about the Subject Tests and what they're used for. We'll explain the differences between SAT II: Mathematics Level IC and

SAT II: Mathematics Level IIC and help you decide which test is the one for you. And then we'll provide you with the study plan that's just right for the test that you're preparing for and the amount of time you have to devote to preparation.

Review Facts, Formulas, and Strategies

Once you have the big picture, it's time to focus on the math that's tested. The second section of this book, "Facts, Formulas, and Strategies," gives you a complete tour of all the typical types of questions that appear on these tests, and a succinct review of the math you need to know to answer such questions. Each chapter in this section begins with a diagnostic test. If your time is limited, you can use the diagnostic tests to bypass the material you already know well enough and to zero in on what you need to work on. Each of these chapters also ends with a follow-up test. When you feel you have mastered the material in a chapter, take the follow-up test to make sure.

Finding Your Way

How you use this book depends on which test you're taking and how much time you have. Let's take a look at five typical students who are planning to take the SAT II: Mathematics Subject Tests. Our five hypothetical students could use this book in five different ways. Which student's study plan best matches your situation?

 "I'm taking Level IC a month from today."

Angela has plenty of time to prep for Level IC. If you're like Angela, and you have at least two weeks to prepare, then we recommend that you do everything in this book that relates to the test you're taking.

 "I'm taking Level IC in a week."

If you're like Bill, you'll need a shortcut. If you have fewer than two weeks but more than two days to prepare, then we recommend that you use the diagnostic tests to determine which chapters you can safely skim, or even skip. That's the shortcut for Level IC. The chapters that review content (chapters 4–10) will guide you to the shortcut.

"I'm taking Level IIC six weeks from today."

Carey has plenty of time to prep for Level IIC. If you're like Carey, and you have at least two weeks to prepare, then we recommend that you do everything in this book that relates to the test you're taking. That's the standard plan for Level IIC.

 "I'm taking Level IIC six days from today."

David needs to quickly brush up on some weak spots to get ready for the Level IIC Test. If you have fewer than two weeks but more than two days to prepare, then we recommend that you use the diagnostic tests to determine which chapters you can safely skim, or even skip. The chapters that review content (chapters 4–10) will guide you to the shortcut.

 "Help! It's two days before Test Day!"

Eric is in a panic. You don't need to freak, even if you're in Eric's situation. Use our Panic Plan to get through this book. If you have only a day or two to prepare for the test, then you don't have time to prepare thoroughly. But that doesn't mean you should just give up and not prepare at all. There's still a lot you can do to improve your potential score. First and foremost, you should become familiar with the test. Read the first three chapters in this book. And if you do nothing else, you should at least sit down and work through one of the full-length practice tests at the back of this book under reasonably testlike conditions.

When you finish the practice test, check your answers and look at the explanations to the questions you didn't get right. When you come across a topic that you only half remember, turn to the appropriate chapter in this book for a quick review. When you come across a topic you don't remember or understand at all, skip it. You don't have time to learn and assimilate completely new material. At least you'll know to skip any similar question you might encounter on the actual SAT II: Mathematics Subject Test.

The Icons

As you work your way through this book, you'll see the following helpful icons used repeatedly. Here's what they mean.

Finding Your Way This icon appears next to information that helps you use the chapters to fit your study plan.

 Facts and Formulas This icon highlights facts and formulas that you need to know for the SAT II: Mathematics Test.

 Kaplan Strategies This icon indicates a Kaplan test-taking strategy that can help boost your score.

 Test Topics This icon appears next to each discussion of a math topic that's tested on the SAT II.

 Level IIC This icon appears next to more difficult questions that would appear primarily on the Level IIC Test.

College Board Publications

The College Board has released some test questions that you might want to look at. The free pamphlet *Taking the SAT II: Subject Tests* has a few sample questions. Even better, the College Board's book *The Official Guide to SAT II: Subject Tests* has full-length, in-format tests.

Take the Full-Length Practice Tests

At the back of the book are two full-length Level IC practice tests and two full-length Level IIC practice tests. The best way to use these tests is to take them under testlike conditions. Don't just drop in and do a random question here and there. Use these tests to gain experience with the complete testing experience, including pacing and endurance. You can do these tests at any time. You don't have to save them all until after you've read this whole book. You might want to use one of the tests as a diagnostic; just be sure to save at least one test for your dress rehearsal sometime in the last week before Test Day.

Take a Break Before Test Day

If possible, don't study the night before the test. Relax! Read a book or watch a movie. Get a good night's sleep. Eat a light breakfast the morning of the test and quickly review a few questions if you feel like it (just enough to get your mind on the test). Walk into the test center with confidence—you're ready!

A Special Note for International Students

Approximately 500,000 international students pursued academic degrees at the undergraduate, graduate, or professional school level at U.S. universities during the 1995–1996 academic year, according to the Institute of International Education's *Open Doors* report. Almost 50 percent of these students were studying for a bachelor's or first university degree. This trend of pursuing higher education in the United States is expected to continue well into the next century. Business, management, engineering, and the physical and life sciences are particularly popular majors for students coming to the United States from other countries.

If you are not from the United States, but are considering attending a U.S. college or university, here's what you'll need to get started:

- If English is not your first language, start there. You'll probably need to take the Test of English as a Foreign Language (TOEFL) and the Test of Written English (TWE), or show some other evidence that you are fully proficient in English in order to complete an academic degree program. Colleges and universities in the United States will differ on what they consider to be an acceptable TOEFL score. A minimum TOEFL score of 550 or better is often expected by the more prestigious and competitive institutions. Because American undergraduate programs require all students to take a certain number of general education courses, all students, even math and computer science students, need to be able to communicate well in spoken and written English.

  You might also need to take the Scholastic Assessment Test (SAT) or the American College Test (ACT). Many undergraduate institutions in the United States require both the SAT and TOEFL of international students.

- There are over 2,700 accredited colleges and universities in the United States, so selecting the correct undergraduate school can be a confusing task for anyone. You will need to get help from a good advisor or at least a good college guide that explains the different types of programs and gives you some information on

Applying to Access America

To get more information, or to apply for admission to any of Kaplan's programs for international students or professionals, you can write to us at:

Kaplan Educational Centers
International Admissions Dept.
888 Seventh Avenue
New York, NY 10106

Phone: (800) 522-7770 or
01-212-262-4980
outside the United States

Fax: 01-212-957-1654

E-mail: world@kaplan.com

Internet: http://www.kaplan.com/intl

how to choose wisely. Since admission to many undergraduate programs is quite competitive, you might also want to select three or four colleges and complete applications for each school.

- You should begin the application process at least a year in advance. An increasing number of schools accept applications year round. In any case, find out the application deadlines and plan accordingly. Although September (the fall semester) is the traditional time to begin university study in the United States, at most schools you can also enter in January (the spring semester).

- Finally, you will need to obtain an I-20 Certificate of Eligibility in order to obtain an F-1 Student Visa to study in the United States. This you will request from the university. The school will send you the I-20 document once you have been accepted.

For an overview of the undergraduate admissions process, see the appendix on college admissions in this book. For details about the admissions requirements, curriculum, and other vital information on top colleges and universities, see Kaplan's *Road to College*.

Access America™

If you need more help with the complex process of undergraduate school admissions and information about the variety of programs available, you might be interested in Kaplan's Access America program.

Kaplan created Access America to assist students and professionals from outside the United States who want to enter the U.S. university system. The program was designed for students who have received the bulk of their primary and secondary education outside the United States in a language other than English. Access America also has programs for obtaining professional certification in the United States. Here's a brief description of some of the help available through Access America.

The TOEFL Plus Program

At the heart of the Access America program is the intensive TOEFL Plus Academic English program. This comprehensive English course prepares students to achieve a high level of proficiency in English in order to successfully complete an academic degree. The TOEFL Plus course combines personalized instruction with guided self-study to help students gain this proficiency in a short time. Certificates of Achievement in English are awarded to certify each student's level of proficiency.

Undergraduate School/SAT Preparation

If your goal is to complete a bachelor of arts (B.A.) or bachelor of science (B.S.) degree in the United States, Kaplan will help you prepare for the SAT or ACT, while helping you understand the American system of education.

Stress Management

Don't freak out about the SAT II: Mathematics Subject Test. This section contains stress-relief tips to help you perform your best on Test Day.

"She says the lava lamp relaxes her during the test."

The Kaplan Advantage™

The countdown has begun. Your date with THE TEST is looming on the horizon. Anxiety is on the rise. The butterflies in your stomach have gone ballistic. Perhaps you feel as if the last thing you ate has turned into a lead ball. Your thinking is getting cloudy. Maybe you think you won't be ready. Maybe you already know your stuff, but you're going into panic mode anyway. Worst of all, you're not sure of what to do about it.

Don't freak! It is possible to tame that anxiety and stress—before *and* during the test. We'll show you how. You won't believe how quickly and easily you can deal with that killer anxiety.

Make the Most of Your Prep Time

Lack of control is one of the prime causes of stress. A ton of research shows that if you don't have a sense of control over what's happening in your life, you can easily end up feeling helpless and hopeless. So, just having concrete things to do and to think about—taking control—will help reduce your stress. This section shows you how to take control during the days leading up to taking the test.

Identify the Sources of Stress

In the space provided, jot down (in pencil) anything you identify as a source of your test-related stress. The idea is to pin down that free-floating anxiety so that you can take control of it. Here are some common examples to get you started:

- I always freeze up on tests.
- I'm nervous about trig (or functions, or coordinate geometry, etcetera).
- I need a good/great score to go to Acme College.

Avoid Must-y Thinking

Let go of "must-y" thoughts, those notions that you must do something in a certain way—for example, "I must get a great score, or else!" "I must meet Mom and Dad's expectations."

- My older brother/sister/best friend/girl- or boyfriend did really well. I *must* match their scores or do better.
- My parents, who are paying for school, will be really disappointed if I don't test well.
- I'm afraid of losing my focus and concentration.
- I'm afraid I'm not spending enough time preparing.
- I study like crazy, but nothing seems to stick in my mind.
- I always run out of time and get panicky.
- I feel as though thinking is becoming like wading through thick mud.

Sources of Stress

I can't concentrate during tests.

I want to get the same score or better as two of my friends

Take a few minutes to think about the things you've just written down. Then rewrite them in some sort of order. List the statements you most associate with your stress and anxiety first, and put the least disturbing items last. Chances are, the top of the list is a fairly accurate description of exactly how you react to test anxiety, both physically and mentally. The later items usually describe your fears (disappointing Mom and Dad, looking bad, etcetera). As you write the list, you're forming a hierarchy of items so you can deal first with the anxiety provokers that bug you most. Very often, taking care of the major items from the top of the list goes a long way toward relieving overall testing anxiety. You probably won't have to bother with the stuff you placed last.

Take Stock of Your Strengths and Weaknesses

Take one minute to list the areas of the test that you are good at. They can be general ("algebra") or specific ("quadratic equations"). Put down as many as you can think of, and if possible, time yourself. Write for the entire time; don't stop writing until you've reached the one-minute stopping point.

Strong Test Subjects

_____   _____

_____   _____

_____   _____

_____   _____

_____   _____

Next, take one minute to list areas of the test you're not so good at, just plain bad at, have failed at, or keep failing at. Again, keep it to one minute, and continue writing until you reach the cutoff. Don't be afraid to identify and write down your weak spots! In all probability, as you do both lists, you'll find you are strong in some areas and not so strong in others. Taking stock of your assets _and_ liabilities lets you know the areas you don't have to worry about, and the ones that will demand extra attention and effort.

Weak Test Subjects

_____   _____

_____   _____

_____   _____

_____   _____

_____   _____

Now, go back to the "good" list, and expand it for two minutes. Take the general items on that first list and make them more specific; take the specific items and expand them into more general conclusions. Naturally, if anything new comes to mind, jot it down. Focus all of your attention and effort on your strengths. Don't underestimate yourself or your abilities. Give yourself full credit. At the same time, don't list strengths you don't really have; you'll only be fooling yourself.

Expanding from general to specific might go as follows. If you listed "algebra" as a broad topic you feel strong in, you would then narrow your focus to include areas of this subject about which you are particularly knowledgeable. Your areas of strength might include multiplying polynomials, working with exponents, factoring, solving simultaneous equations, etcetera.

Whatever you know comfortably goes on your "good" list. Okay. You've got the picture. Now, get ready, check your starting time, and start writing down items on your expanded "good" list.

Very Superstitious

Stress expert Stephen Sideroff, Ph.D., tells of a client who always stressed out before, during, and even after taking tests. Yet, she always got outstanding scores. It became obvious that she was thinking superstitiously—subconsciously believing that the great scores were a result of her worrying. She didn't trust herself, and believed that if she didn't worry she wouldn't study hard enough. Sideroff convinced her to take a risk and work on relaxing before her next test. She did, and her test results were still as good as ever—which broke her cycle of superstitious thinking.

Stress Tip

Don't work in a messy or cramped area. Before you sit down to study, clear yourself a nice, open space. And, make sure you have books, paper, pencils—whatever tools you will need—within easy reach before you sit down to study.

Link Your Thoughts

When you're committing new information to memory, link one fact to another, much as elephants are linked trunk to tail in a circus parade. Visualize an image (preferably a bizarre one) that connects the thoughts. You'll remember them in the same linked way, with one thought easily bringing the next to your mind.

Don't Force It

Never try to force relaxation. You'll only get frustrated and find yourself even more uptight. Be passive.

Strong Test Subjects: An Expanded List

_____   _____

_____   _____

_____   _____

_____   _____

_____   _____

After you've stopped, check your time. Did you find yourself going beyond the two minutes allotted? Did you write down more things than you thought you knew? Is it possible you know more than you've given yourself credit for? Could that mean you've found a number of areas in which you feel strong?

You just took an active step toward helping yourself. Notice any increased feelings of confidence? Enjoy them.

Here's another way to think about your writing exercise. Every area of strength and confidence you can identify is much like having a reserve of solid gold at Fort Knox. You'll be able to draw on your reserves as you need them. You can use your reserves to solve difficult questions, maintain confidence, and keep test stress and anxiety at a distance. The encouraging thing is that every time you recognize another area of strength, succeed at coming up with a solution, or get a good score on a test, you increase your reserves. And, there is absolutely no limit to how much self-confidence you can have or how good you can feel about yourself.

Imagine Yourself Succeeding

This next little group of exercises is both physical and mental. It's a natural follow-up to what you've just accomplished with your lists.

First, get yourself into a comfortable sitting position in a quiet setting. Wear loose clothes. If you wear glasses, take them off. Then, close your eyes and breathe in a deep, satisfying breath of air. Really fill your lungs until your rib cage is fully expanded and you can't take in any more. Then, exhale the air completely. Imagine you're blowing out a candle with your last little puff of air. Do this two or three more times, filling your lungs to their maximum and emptying them totally. Keep your eyes closed, comfortably but not tightly. Let your body sink deeper into the chair as you become even more comfortable.

With your eyes shut you can notice something very interesting. You're no longer dealing with the worrisome stuff going on in the world *outside* of you. Now you can concentrate on what happens *inside* you. The more you recognize your own physical reactions to stress and anxiety, the more you can do about them. You might not realize it, but you've begun to regain a sense of being in control.

Let images begin to form on the "viewing screens" on the back of your eyelids. You're experiencing visualizations from the place in your mind that makes pictures. Allow the images to come easily and naturally; don't force them. Imagine yourself in a relaxing situation. It might be in a special place you've visited before or one you've read about. It can be a fictional location that you create in your imagination, but a real-life memory of a place or situation you know is usually better. Make it as detailed as possible, and notice as much as you can.

If you don't see this relaxing place sharply or in living color, it doesn't mean the exercise won't work for you. Some people can visualize in great detail, while others get only a sense of an image. What's important is not how sharp the details or colors, but how well you're able to manipulate the images. If you can conjure up finely detailed images, great. If you have only a faint sense of the images, that's okay—you'll still experience all the benefits of the exercise.

Think about the sights, the sounds, the smells, even the tastes and textures associated with your relaxing situation. *See* and *feel* yourself in this special place. Say your special place is the beach, for example. Feel how warm the sand is. Are you lying on a blanket, or sitting up and looking out at the water? Hear the waves hitting the shore, and the occasional seagull. Feel a comfortable breeze. If your special place is a garden or park, look up and see the way sunlight filters through the trees. Smell your favorite flowers. Hear some chimes gently playing and birds chirping.

Stay focused on the images as you sink farther back into your chair. Breathe easily and naturally. You might have the sensations of any stress or tension draining from your muscles and flowing downward, out your feet and away from you.

Take a moment to check how you're feeling. Notice how comfortable you've become. Imagine how much easier it would be if you could take the test feeling this relaxed and in this state of ease. You've coupled the images of your special place with sensations of comfort and relaxation. You've also found a way to become relaxed simply by visualizing your own safe, special place.

Now, close your eyes and start remembering a real-life situation in which you did well on a test. If you can't come up with one, remember a situation in which you did something (academic or otherwise) that you were really proud of—a genuine accomplishment. Make the memory as detailed as possible. Think about the sights, the sounds, the smells, even the tastes associated with this remembered experience. Remember how confident you felt as you accomplished your goal. Now start thinking about the upcoming test. Keep your thoughts and feelings in line with that successful

Ocean Dumping

Visualize a beautiful beach, with white sand, blue skies, sparkling water, a warm sun, and seagulls. See yourself walking on the beach, carrying a small plastic pail. Stop at a good spot and put your worries and whatever may be bugging you into the pail. Drop it at the water's edge and watch it drift out to sea. When the pail is out of sight, walk on.

The "New Age" of Relaxation

Here are some more tips for beating stress:

- Massage, especially shiatsu—see if it's offered through your school's phys ed department, or at the local "Y."

- Check out a book on acupressure, and find those points on your body where you can press a "relax button."

- If you're especially sensitive to smells, you might want to try some aromatherapy. Lavender oil, for example, is said to have relaxing properties. Health food stores, drug stores, and New Age bookstores may carry aromatherapy oils.

- Many health food stores carry herbs and supplements that have relaxing properties, and they often have a specialist on staff who can tell you about them.

Stress Tip

Don't forget that your school probably has counseling available. If you can't conquer test stress on your own, make an appointment at the counseling center. That's what counselors are there for.

experience. Don't make comparisons between them. Just imagine taking the upcoming test with the same feelings of confidence and relaxed control.

This exercise is a great way to bring the test down to Earth. You should practice this exercise often, especially when the prospect of taking the exam starts to bum you out. The more you practice it, the more effective the exercise will be for you.

What Do You Want to Accomplish in the Time Remaining?

The whole point to this next exercise is sort of like checking out a used car you might want to buy. You'd want to know up front what the car's weak points are, right? Knowing that influences your whole shopping-for-a-used-car campaign. So it is with your conquering-test-stress campaign: Knowing what your weak points are ahead of time helps you prepare.

So let's get back to the list of your weak points. Take two minutes to expand it just as you did with your "good" list. Be honest with yourself without going overboard. It's an accurate appraisal of the test areas that give you trouble. So, pick up your pencil, check the clock, and start writing.

Weak Test Subjects: An Expanded List

_____   _____

_____   _____

_____   _____

_____   _____

_____   _____

How did you do? Were you able to keep writing for the full two minutes? Has making this "weak" list helped you become more clear about the specific areas you need to address?

Facing your weak spots gives you some distinct advantages. It helps a lot to find out where you need to spend extra effort. Increased exposure to tough material makes it more familiar and less intimidating. (After all, we mostly fear what we don't know and are probably afraid to face.) You'll feel better about yourself because you're dealing directly with areas of the test that bring on your anxiety. You can't help feeling more confident when you know you're actively strengthening your chances of earning a higher overall test score.

Exercise Your Frustrations Away

Whether it is jogging, walking, biking, mild aerobics, pushups, or a pickup basketball game, physical exercise is a very effective way to stimulate both your mind and body and to improve your ability to think and concentrate. A surprising number of students get out of the habit of regular exercise, ironically because they're spending so much time prepping for exams. Also, sedentary people—this is a medical fact—get less oxygen to the blood and hence to the head than active people. You can live fine with a little less oxygen; you just can't think as well.

Any big test is a bit like a race. Thinking clearly at the end is just as important as having a quick mind early on. If you can't sustain your energy level in the last sections of the exam, there's too good a chance you could blow it. You need a fit body that can weather the demands any big exam puts on you. Along with a good diet and adequate sleep, exercise is an important part of keeping yourself in fighting shape and thinking clearly for the long haul.

There's another thing that happens when students don't make exercise an integral part of their test preparation. Like any organism in nature, you operate best if all your "energy systems" are in balance. Studying uses a lot of energy, but it's all mental. When you take a study break, do something active instead of raiding the fridge or vegging out in front of the TV. Take a 5- to 10-minute activity break for every 50 or 60 minutes that you study. The physical exertion gets your body into the act, which helps to keep your mind and body in sync. Then, when you finish studying for the night and hit the sack, you won't lie there, tense and unable to sleep because your head is overtired and your body wants to pump iron or run a marathon.

One warning about exercise, however: It's not a good idea to exercise vigorously right before you go to bed. This could easily cause sleep onset problems. For the same reason, it's also not a good idea to study right up to bedtime. Make time for a "buffer period" before you go to bed: For 30 to 60 minutes, just take a hot shower, meditate, simply veg out.

Stress Tip

If you want to play music, keep it low and in the background. Music with a regular, mathematical rhythm—reggae, for example—aids the learning process. A recording of ocean waves is also soothing.

Take a Hike, Pal

When you're in the middle of studying and hit a wall, take a short, brisk walk. Breathe deeply and swing your arms as you walk. Clear your mind. (And, don't forget to look for flowers that grow in the cracks of the sidewalk.)

Get High . . . Naturally

Exercise can give you a natural high, which is the only kind of high you should be aiming for. Using drugs (prescription or recreational) specifically to prepare for and take a big test is definitely self-defeating. (And if they're illegal drugs, you can end up with a bigger problem than the SAT on your hands.) Except for the drugs that occur naturally in your brain, *every* drug has major drawbacks—and a false sense of security is only one of them.

Cyberstress

If you spend a lot of time in cyberspace anyway, do a search for the phrase "stress management." There's a ton of stress advice on the Net, including material specifically for students.

Nutrition and Stress: The Dos and Don'ts

Do eat:

- Fruits and vegetables (raw is best, or just lightly steamed or nuked)

- Low-fat protein such as fish, skinless poultry, beans, and legumes (like lentils)

- Whole grains such as brown rice, whole wheat bread, and pastas (no bleached flour)

Don't eat:

- Refined sugar; sweet, high-fat snacks (simple carbohydrates like sugar make stress worse and fatty foods lower your immunity)

- Salty foods (they can deplete potassium, which you need for nerve functions)

You may have heard that popping uppers helps you study by keeping you alert. If they're illegal, definitely forget about it. They wouldn't really work anyway, since amphetamines make it hard to retain information. Mild stimulants, such as coffee, cola, or over-the-counter caffeine pills can sometimes help as you study, since they keep you alert. On the down side, they can also lead to agitation, restlessness, and insomnia. Some people can drink a pot of high-octane coffee and sleep like a baby. Others have one cup and start to vibrate. It all depends on your tolerance for caffeine. Remember, a little anxiety is a good thing. The adrenaline that gets pumped into your bloodstream helps you stay alert and think more clearly. But, too much anxiety and you can't think straight at all.

Alcohol and other depressants are out, too. Again, if it's illegal, forget about it. Depressants wouldn't work anyway, since they lead to the inevitable hangover/crash, fuzzy thinking, and lousy sense of judgment. These would not help you ace the test.

Instead, go for endorphins—the "natural morphine." Endorphins have no side effects and they're free—you've already got them in your brain. It just takes some exercise to release them. Running around on the basketball court, bicycling, swimming, aerobics, power walking—these activities cause endorphins to occupy certain spots in your brain's neural synapses. In addition, exercise develops staying power and increases the oxygen transfer to your brain. Go into the test naturally.

Take a Deep Breath . . .

Here's another natural route to relaxation and invigoration. It's a classic isometric exercise that you can do whenever you get stressed out—just before the test begins, even *during* the test. It's very simple and takes just a few minutes.

Close your eyes. Starting with your eyes and—*without holding your breath*—gradually tighten every muscle in your body (but not to the point of pain) in the following sequence:

1. Close your eyes tightly.
2. Squeeze your nose and mouth together so that your whole face is scrunched up. (If it makes you self-conscious to do this in the test room, skip the face-scrunching part.)
3. Pull your chin into your chest, and pull your shoulders together.
4. Tighten your arms to your body, then clench your hands into tight fists.
5. Pull in your stomach.
6. Squeeze your thighs and buttocks together, and tighten your calves.

7. Stretch your feet, then curl your toes (watch out for cramping in this part).

At this point, every muscle should be tightened. Now, relax your body, one part at a time, *in reverse order*, starting with your toes. Let the tension drop out of each muscle. The entire process might take five minutes from start to finish (maybe a couple of minutes during the test). This clenching and unclenching exercise should help you to feel very relaxed.

And Keep Breathing

Conscious attention to breathing is an excellent way of managing test stress (or any stress, for that matter). The majority of people who get into trouble during tests take shallow breaths. They breathe using only their upper chests and shoulder muscles, and may even hold their breath for long periods of time. Conversely, the test taker who by accident or design keeps breathing normally and rhythmically is likely to be more relaxed and in better control during the entire test experience.

So, now is the time to get into the habit of relaxed breathing. Do the next exercise to learn to breathe in a natural, easy rhythm. By the way, this is another technique you can use during the test to collect your thoughts and ward off excess stress. The entire exercise should take no more than three to five minutes.

With your eyes still closed, breathe in slowly and *deeply* through your nose. Hold the breath for a bit, and then release it through your mouth. The key is to breathe slowly and deeply by using your diaphragm (the big band of muscle that spans your body just above your waist) to draw air in and out naturally and effortlessly. Breathing with your diaphragm encourages relaxation and helps minimize tension.

As you breathe, imagine that colored air is flowing into your lungs. Choose any color you like, from a single color to a rainbow. With each breath, the air fills your body from the top of your head to the tips of your toes. Continue inhaling the colored air until it occupies every part of you, bones and muscles included. Once you have completely filled yourself with the colored air, picture an opening somewhere on your body, either natural or imagined. Now, with each breath you exhale, some of the colored air will pass out the opening and leave your body. The level of the air (much like the water in a glass as it is emptied) will begin to drop. It will descend progressively lower, from your head down to your feet. As you continue to exhale the colored air, watch the level go lower and lower, farther and farther down your body. As the last of the colored air passes out of the opening, the level will drop down to your toes and disappear. Stay quiet for just a moment. Then notice how relaxed and comfortable you feel.

The Relaxation Paradox

Forcing relaxation is like asking yourself to flap your arms and fly. You can't do it, and every push and prod only gets you more frustrated. Relaxation is something you don't work at. You simply let it happen. Think about it. When was the last time you tried to force yourself to go to sleep, and it worked?

Stress Tip

A lamp with a 75-watt bulb is optimal for studying. But don't put it so close to your study material that you create a glare.

Stress Tip

Don't study on your bed, especially if you have problems with insomnia. Your mind might start to associate the bed with work, and make it even harder for you to fall asleep.

Dress for Success

On the day of the test, wear loose layers. That way, you'll be prepared no matter what the temperature of the room is. (An uncomfortable temperature will just distract you from the job at hand.)

And, if you have an item of clothing that you tend to feel "lucky" or confident in—a shirt, a pair of jeans, whatever—wear it. A little totem couldn't hurt.

Quick Tips for the Days Just Before the Exam

- The best test takers do less and less as the test approaches. Taper off your study schedule and take it easy on yourself. You want to be relaxed and ready on the day of the test. Give yourself time off, especially the evening before the exam. By then, if you've studied well, everything you need to know is firmly stored in your memory banks.

- Positive self-talk can be extremely liberating and invigorating, especially as the test looms closer. Tell yourself things such as, "I choose to take this test" rather than "I have to"; "I will do well" rather than "I hope things go well"; "I can" rather than "I cannot." Be aware of negative, self-defeating thoughts and images and immediately counter any you become aware of. Replace them with affirming statements that encourage your self-esteem and confidence. Create and practice visualizations that build on your positive statements.

- Get your act together sooner rather than later. Have everything (including choice of clothing) laid out days in advance. Most important, know where the test will be held and the easiest, quickest way to get there. You will gain great peace of mind if you know that all the little details—gas in the car, directions, etcetera—are firmly in your control before the day of the test.

- Experience the test site a few days in advance. This is very helpful if you are especially anxious. If at all possible, find out what room your part of the alphabet is assigned to, and try to sit there (by yourself) for a while. Better yet, bring some practice material and do at least a section or two, if not an entire practice test, in that room. In this situation, familiarity doesn't breed contempt, it generates comfort and confidence.

- Forego any practice on the day before the test. It's in your best interest to marshal your physical and psychological resources for 24 hours or so. Even race horses are kept in the paddock and treated like princes the day before a race. Keep the upcoming test out of your consciousness; go to a movie, take a pleasant hike, or just relax. Don't eat junk food or tons of sugar. And—of course—get plenty of rest the night before. Just don't go to bed too early. It's hard to fall asleep earlier than you're used to, and you don't want to lie there thinking about the test.

Thumbs Up for Meditation

Once relegated to the fringes of the medical world, meditation, biofeed-back, and hypnosis are increasingly recommended by medical researchers to reduce pain from headaches, back problems—even cancer. Think of what these powerful techniques could do for your test-related stress and anxiety.

Effective meditation is based primarily on two relaxation methods you've already learned: body awareness and breathing. A couple of different meditation techniques follow. Experience them both, and choose the one that works best for you.

Breath Meditation

Make yourself comfortable, either sitting or lying down. For this meditation you can keep your eyes opened or closed. You're going to concentrate on your breathing. The goal of the meditation is to notice everything you can about your breath as it enters and leaves your body. Take three to five breaths each time you practice the meditation; this set of breaths should take about a minute to complete.

Take a deep breath and hold it for 5 to 10 seconds. When you exhale, let the breath out very slowly. Feel the tension flowing out of you along with the breath that leaves your body. Pay close attention to the air as it flows in and out of your nostrils. Observe how cool it is as you inhale and how warm your breath is when you exhale. As you expel the air, say to yourself a cue word such as *calm* or *relax*. Once you've exhaled all the air from your lungs, start the next long, slow inhale. Notice how relaxed feelings increase as you slowly exhale and again hear your cue words.

Mantra Meditation

For this type of meditation experience you'll need a mental device (a mantra), a passive attitude (don't *try* to do anything), and a position in which you can be comfortable. You're going to focus your total attention on a mantra you create. It should be emotionally neutral, repetitive, and monotonous, and your aim is to fully occupy your mind with it. Furthermore, you want to do the meditation passively, with no goal in your head of how relaxed you're supposed to be. This is a great way to prepare for studying or taking the test. It clears your head of extraneous thoughts and gets you focused and at ease.

Sit comfortably and close your eyes. Begin to relax by letting your body go limp. Create a relaxed mental attitude and know there's no need for you to force anything. You're simply going to let something happen. Breathe through your nose. Take calm, easy breaths and as you exhale, say your mantra (*one, ohhm, aah, soup*—whatever is emotionally neutral for you) to yourself. Repeat the mantra each time you breathe out. Let feel-

Breathe Like a Baby

A baby or young child is the best model for demonstrating how to breathe most efficiently and comfortably. Only its stomach moves as it inhales and exhales. The action is virtually effortless.

Think Good Thoughts

Create a set of positive but brief affirmations and mentally repeat them to yourself just before you fall asleep at night. (That's when your mind is very open to suggestion.) You'll find yourself feeling a lot more positive in the morning.

Periodically repeating your affirmations during the day makes them more effective.

ings of relaxation grow as you focus on the mantra and your slow breathing. Don't worry if your mind wanders. Simply return to the mantra and continue letting go. Experience this meditation for 10 to 15 minutes.

Handling Stress During the Test

The biggest stress monster will be the test itself. Fear not; there are methods of quelling your stress during the test.

- Keep moving forward instead of getting bogged down in a difficult question. You don't have to get everything right to achieve a fine score. The best test takers skip difficult material temporarily in search of the easier stuff. They mark the ones that require extra time and thought. This strategy buys time and builds confidence so you can handle the tough stuff later.

- Don't be thrown if other test takers seem to be working more furiously than you are. Continue to spend your time patiently thinking through your answers; it's going to lead better results. Don't mistake the other people's sheer activity as signs of progress and higher scores.

- *Keep breathing!* Weak test takers tend to forget to breathe properly as the test proceeds. They start holding their breath without realizing it, or they breathe erratically or arrhythmically. Improper breathing interferes with clear thinking.

- Some quick isometrics during the test—especially if concentration is wandering or energy is waning—can help. Try this: Put your palms together and press intensely for a few seconds. Concentrate on the tension you feel through your palms, wrists, forearms, and up into your biceps and shoulders. Then, quickly release the pressure. Feel the difference as you let go. Focus on the warm relaxation that floods through the muscles. Now you're ready to return to the task.

- Here's another isometric that will relieve tension in both your neck and eye muscles. Slowly rotate your head from side to side, turning your head and eyes to look as far back over each shoulder as you can. Feel the muscles stretch on one side of your neck as they contract on the other. Repeat five times in each direction.

    With what you've just learned here, you're armed and ready to do battle with the test. This book and your studies will give you the information you'll need to answer the questions. It's all firmly planted in your mind. You also know how to deal with any excess tension that might come along, both when you're studying for and taking the exam. You've experienced everything you need to tame your test anxiety and stress. You're going to get a great score.

What Are "Signs of a Winner," Alex?

Here's some advice from a Kaplan instructor who won big on *Jeopardy!*™ In the green room before the show, he noticed that the contestants who were quiet and "within themselves" were the ones who did great on the show. The contestants who did not perform as well were the ones who were fact-cramming, talking a lot, and generally being manic before the show. Lesson: Spend the final hours leading up to the test getting sleep, meditating, and generally relaxing.

KAPLAN

The Basics

In this section, we'll cover the basics. We'll provide you with an overview of the SAT II: Subject Tests, tell you what's tested on the two levels of the SAT II: Mathematics Subject Test, and give you some important calculator tips.

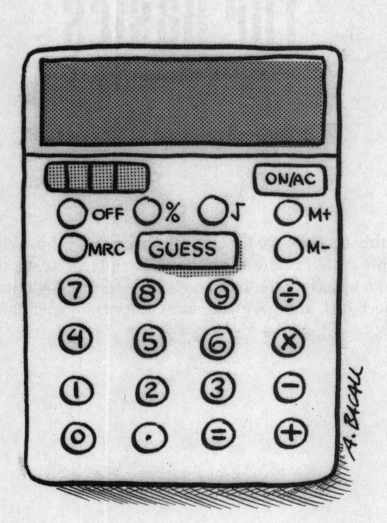

CHAPTER 1

Getting Ready for SAT II: Mathematics

You're serious about going to the college of your choice. You wouldn't have opened this book otherwise. You've made a wise choice, because this book can help you to achieve your goal. It'll show you how to score your best on the SAT II: Mathematics Subject Test. But before turning to the math content, let's look at the SAT II as a whole.

Frequently Asked Questions About the SAT II

The following background information about the SAT II is important to keep in mind as you get ready to prep for the SAT II: Math Subject Tests.

What Is the SAT II?

Known until 1994 as the College Board Achievement Tests, the SAT II is actually a set of more than 20 different Subject Tests. These tests are designed to measure what you have learned in such subjects as Literature, American History and Social Studies, Biology, and Spanish. Each test lasts one hour and consists entirely of multiple-choice questions, except for the Writing Test, which has a 20-minute essay section and a 40-minute multiple-choice section. On any one test date, you can take one, two, or three Subject Tests.

How Does the SAT II Differ from the SAT I?

SAT I is largely a test of verbal and math skills. True, you need to know some vocabulary and some formulas for the SAT I; but it's designed to measure how well you read and think rather than how much you remember. The SAT II tests are very different. They're designed to measure what you know about specific disciplines. Sure, critical reading and thinking skills play a part on these tests, but their main purpose is to determine exactly what you know about writing, math, history, chemistry, and so on.

S.A.T.
What's That Spell?

SAT stands for "Scholastic Assessment Test."

Well, What Do You Know?

The SAT II tests are intended to find out what you know about specific subjects.

Dual Role

Colleges use your SAT II scores in both admissions and placement decisions.

Call Your Colleges

Many colleges require you to take certain SAT II tests. Check with all of the schools you're interested in applying to before deciding on which tests to take.

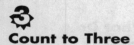

Count to Three

You can take up to three SAT II Tests in one day. The Writing Test must be taken first.

How Do Colleges Use the SAT II?

Many people will tell you that the SATs (I and II alike) measure only your ability to perform on standardized exams—that they measure neither your reading and thinking skills nor your level of knowledge. Maybe they're right. But these people don't work for colleges. Those schools that require SATs feel that they are an important indicator of your ability to succeed in college. Specifically, they use your scores in one or both of two ways: to help them make admissions and/or placement decisions.

Like the SAT I, the SAT II tests provide schools with a standard measure of academic performance, which they use to compare you to applicants from different high schools and different educational backgrounds. This information helps them to decide whether you're ready to handle their curriculum.

SAT II scores may also be used to decide what course of study is appropriate for you once you've been admitted. A low score on the Writing Test, for example, might mean that you have to take a remedial English course. Conversely, a high score on an SAT II: Mathematics Test may mean that you'll be exempted from an introductory math course.

Which SAT II Tests Should I Take?

The simple answer is: those that you'll do well on. High scores, after all, can only help your chances for admission. Unfortunately, many colleges demand that you take particular tests, usually the Writing Test and/or one of the Mathematics Tests. Some schools will give you some choice in the matter, especially if they want you to take a total of three tests. Before you register to take any tests, therefore, check with the colleges you're interested in to find out exactly which tests they require. Don't rely on high school guidance counselors or admissions handbooks for this information. They might not give you accurate or current information.

When Are the SAT II Tests Administered?

Most of the SAT II Tests are administered six times a year: in October, November, December, January, May, and June. A few of the tests are offered less frequently. Due to admissions deadlines, many colleges insist that you take the SAT II no later than December or January of your senior year in high school. You may even have to take it sooner if you're interested in applying for "early admission" to a school. Those schools that use scores for placement decisions only may allow you to take the SAT II as late as May or June of your senior year. You should check with colleges to find out which test dates are most appropriate for you.

How Do I Register for the SAT II?

The College Board administers the SAT II tests, so you must sign up with them. The easiest way to register is to obtain copies of the *SAT Registration Bulletin* and *Taking the SAT II: Subject Tests*. These publications contain all of the necessary information, including current test dates and fees. They can be obtained at any high school guidance office or directly from the College Board.

You can also register by telephone. If you choose this option, you should still read the College Board publications carefully before you make any decisions.

How Are the SAT II Tests Scored?

Like the SAT I, the SAT II tests are scored on a 200–800 scale.

What's a "Good" Score?

That's tricky. The obvious answer is: the score that the colleges of your choice demand. Keep in mind, though, that SAT II scores are just one piece of information that colleges will use to evaluate you. The decision to accept or reject you will be based on many criteria, including your high school transcript, your SAT I scores, your recommendations, your personal statement, your interview (where applicable), your extracurricular activities, and the like. So, failure to achieve the necessary score doesn't automatically mean that your chances of getting in have been damaged. If you really want a numerical benchmark, a score of 600 is considered very solid.

A College Board service known as Score Choice offers you the chance to see your scores before anyone else. If you're unhappy with a score, you don't have to send it along to colleges. If you decide to take advantage of this service, you'll need to take your SAT II tests well in advance of college deadlines. At the very least, using Score Choice will slow down the reporting process. You may also want to retake one or more tests. Two more points to bear in mind:

- Once you've released a score, it can't be withheld in the future.

- If you use Score Choice, you lose the privilege of having some scores sent to schools for free.

For more information about Score Choice, contact the College Board.

Do the Legwork

Want to register or get more info? You can get copies of the *SAT Registration Bulletin* and *Taking the SAT II: Subject Tests* from the College Board. You can also register for the SAT II by phone.

College Board SAT Program
P.O. Box 6200
Princeton, NJ 08541–6200
(609) 771–7600

Real World

The mean score of the 145,235 students who took the SAT II: Mathematics Level IC Test in 1996 was 570. The mean score of the 76,107 students who took the SAT II: Mathematics Level IIC Test in 1996 was 638.

Source: College Board

Not Like High School

The SAT II tests are very different from the tests you've taken in high school. All questions are worth the same points, and you don't get credit for showing your work.

Pack Your Bag

Gather your test materials the day before the test. You'll need:

- Admission ticket
- Proper form of I.D.
- Some sharpened No. 2 pencils
- Good eraser
- Scientific calculator
- Spare calculator batteries

Don't Get Lost

Learn SAT II directions as you prepare for the tests. You'll have more time to spend answering the questions on Test Day.

What Should I Bring to the SAT II?

It's a good idea to get your test materials together the day before the tests. You'll need an admission ticket; a form of identification (check the *Registration Bulletin* to find out what is and what is not permissible); a few sharpened No. 2 pencils; a good eraser; and a scientific calculator (for Math Level IC or IIC). If you'll be registering as a standby, collect the appropriate forms beforehand. Also, make sure that you know how to get to the test center.

SAT II Mastery

Now that you know a little about the SAT II tests, it's time to let you in on a few basic test-taking skills and strategies that can improve your performance on them. You should practice these skills and strategies as you prepare for the SAT II.

Use the Structure of the Test to Your Advantage.

The SAT II tests are different from the tests that you're used to taking. On your high school tests, you probably go through the questions in order. You probably spend more time on hard questions than on easy ones, since hard questions are generally worth more points. And you often show your work since your teachers tell you that how you approach questions is as important as getting the right answers.

None of this applies to the SAT II tests. You can benefit from moving around within the tests, hard questions are worth the same as easy ones, and it doesn't matter how you answer the questions—only what your answers are.

The SAT II tests are highly predictable. Because the format and directions of the SAT II tests remain unchanged from test to test, you can learn the tests' setups in advance. On Test Day, the various question types on the tests shouldn't be new to you.

One of the easiest things you can do to help your performance on the SAT II tests is to understand the directions before taking the test. Since the instructions are always the same, there's no reason to waste a lot of time on Test Day reading them. Learn them beforehand, as you work through this book and the College Board publications.

SAT II: Mathematics questions are arranged in order of difficulty. Not all of the questions on the SAT II tests are equally difficult. The questions often get harder as you work through different parts of a test. This pattern can work to your benefit. Try to be aware of where you are in a test.

When working on more basic problems, you can generally trust your first impulse—the obvious answer is likely to be correct. As you get to the end of a test section, you need to be a bit more suspicious. Now the answers probably won't come as quickly and easily—if they do, look again because the obvious answers may be wrong. Watch out for answers that just "look right." They may be distractors—wrong answer choices deliberately meant to entice you.

There's no mandatory order to the questions on the SAT II. You're allowed to skip around on the SAT II tests. High scorers know this fact. They move through the tests efficiently. They don't dwell on any one question, even a hard one, until they've tried every question at least once.

When you run into questions that look tough, circle them in your test booklet and skip them for the time being. Go back and try again after you've answered the easier ones if you've got time. After a second look, troublesome questions can turn out to be remarkably simple.

If you've started to answer a question but get confused, quit and go on to the next question. Persistence might pay off in high school, but it usually hurts your SAT II scores. Don't spend so much time answering one hard question that you use up three or four questions' worth of time. That'll cost you points, especially if you don't even get the hard question right.

You can use the so-called guessing penalty to your advantage. You might have heard it said that the SAT II has a "guessing penalty." That's a misnomer. It's really a *wrong-answer penalty*. If you guess wrong, you get a small penalty. If you guess right, you get full credit.

The fact is, if you can eliminate one or more answer choices as definitely wrong, you'll turn the odds in your favor and actually come out ahead by guessing. The fractional points that you lose are meant to offset the points you might get "accidentally" by guessing the correct answer. With practice, however, you'll see that it's often easy to eliminate *several* answer choices on some of the questions.

The answer grid has no heart. It sounds simple, but it's extremely important: Don't make mistakes filling out your answer grid. When time is short, it's easy to get confused going back and forth between your test booklet and your grid. If you know the answers, but misgrid, you won't get the points. Here's how to avoid mistakes.

Always circle the questions you skip. Put a big circle in your test booklet around any question numbers that you skip. When you go back, these questions will be easy to relocate. Also, if you accidentally skip a box on the grid, you'll be able to check your grid against your booklet to see where you went wrong.

There's a Pattern

SAT II: Mathematics questions are arranged in order of difficulty—basic questions first, harder questions last.

Leap Ahead

You should do the questions in the order that's best for you. Don't pass up the opportunity to score easy points by wasting time on hard questions. Skip hard questions until you've gone through every question once. Come back to them later.

Guessing Rule

Don't guess, unless you can eliminate at least one answer choice. Don't leave a question blank, unless you have absolutely no idea about it.

Hit the Spot

A common cause of major SAT II disasters is filling in all of the questions with the right answers—in the wrong spots. Every time you skip a question, circle it in your test booklet and be double sure that you skip it on the answer grid as well.

Think First

Always try to think of the answer to a question before you shop among the answer choices. If you've got some idea of what you're looking for, you'll be less likely to be fooled by "trap" choices.

Always circle the answers you choose. Circling your answers in the test booklet makes it easier to check your grid against your booklet.

Grid five or more answers at once. Don't transfer your answers to the grid after every question. Transfer them after every five questions. That way, you won't keep breaking your concentration to mark the grid. You'll save time and gain accuracy.

Approaching SAT II Questions

Apart from knowing the setup of the SAT II tests that you'll be taking, you've got to have a system for attacking the questions. You wouldn't travel around an unfamiliar city without a map, and you shouldn't approach the SAT II without a plan. What follows is the best method for approaching SAT II questions systematically.

Think about the questions before you look at the answers. The test makers love to put distractors among the answer choices. Distractors are answers that look like they're correct, but aren't. If you jump right into the answer choices without thinking first about what you're looking for, you're much more likely to fall for one of these traps.

Guess—when you can eliminate at least one answer choice. You already know that the "guessing penalty" can work in your favor. Don't simply skip questions that you can't answer. Spend some time with them in order to see whether you can eliminate any of the answer choices. If you can, it pays for you to guess.

Pace yourself. The SAT II tests give you a lot of questions in a short period of time. To get through the tests, you can't spend too much time on any single question. Keep moving through the tests at a good speed. If you run into a hard question, circle it in your test booklet, skip it, and come back to it later if you have time.

You don't have to spend the same amount of time on every question. Ideally, you should be able to work through the easier questions at a brisk, steady clip, and use a little more time on the harder questions. One caution: Don't rush through basic questions just to save time for the harder ones. The basic questions are points in your pocket, and you're better off not getting to some harder questions if it means losing easy points because of careless mistakes. Remember, you don't earn any extra credit for answering hard questions.

Locate quick points if you're running out of time. Some questions can be done more quickly than others because they require less work or because choices can be eliminated more easily. If you start to run out of time, look for these quicker questions.

When you take the SAT II: Subject Tests, you have one clear objective in mind: to score as many points as you can. It's that simple. The rest of this book will show you how to do that on the SAT II: Mathematics Subject Tests.

What's Next

In the next chapter, we'll help you decide whether to take the Level IC Test or the Level IIC Test.

Speed Limit

Work quickly on easier questions to leave more time for harder questions. But not so quickly that you lose points by making careless errors. And it's okay to leave some questions blank if you have to— even if you leave a few blank, you can still get a high score.

CHAPTER 2

Finding Your Level

T he first thing to do to get ready for SAT II: Mathematics is to decide which test you're going to take—Level IC or Level IIC. This chapter will give you the information you need to make that decision: the differences in content, level of difficulty, scoring, and reputation.

Content

The first factor to consider in deciding which test to take is *content*. There's a lot of overlap between what's tested on Level IC and what's tested on Level IIC. But there's also a lot that's tested on Level IIC only, and even some math that's tested on Level IC only.

Level IC is meant to cover the math you'd get in two years of algebra and one year of geometry. Level IIC is meant to cover that much math plus what you'd get in a year of trigonometry and/or precalculus. There is no calculus on either test.

In order to make room for more questions on more advanced topics, Level IIC has fewer questions on the more basic topics. In fact, Level IIC has no plane geometry questions at all. Here's the official breakdown, according to the College Board's publication *Taking the SAT II Subject Tests*.

Approximate Number of Questions by Content Area

Content Area	Level IC	Level IIC
Algebra	15	9
Plane Geometry	10	—
Solid Geometry	3	4
Coordinate Geometry	6	6
Trigonometry	4	10
Functions	6	12
Miscellaneous	6	9

Content at a Glance

Level IC covers two years of algebra and one year of geometry. Level IIC covers two years of algebra, one year of geometry, and one year of trigonometry and/or precalculus. There is no calculus on either test.

Firm Up the Foundations

Don't review math haphazardly. Start with the fundamentals and work your way up to more advanced and esoteric topics.

Level IIC is weighted toward the more advanced topics, but it still tests your understanding of the basics. Take the case of plane geometry. Ostensibly Level IIC has *no* plane geometry questions. But to do a lot of the more advanced Level IIC questions—solid geometry, coordinate geometry, trigonometry—you have to know all about plane geometry.

The Content Pyramid

The seven content areas listed in the chart on the previous page are not discrete and equal areas. Think about how you learned these subjects. You didn't start with trigonometry or functions, did you? Of course not. Math is cumulative. Advanced subjects are built upon basic subjects. Think of the seven content areas as parts of a pyramid.

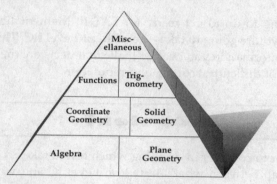

The SAT II: Math Content Pyramid

The content areas on the bottom—algebra and plane geometry—are the foundations upon which all the others are built. When you learned the math represented in this pyramid, you started at the bottom and worked your way up. That's the same way you should review this math in getting ready for the SAT II. Firm up the foundations, and work your way up to more advanced topics.

Now look at the two different question breakdowns in pyramid form.

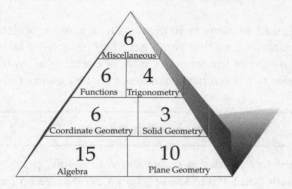

Level IC

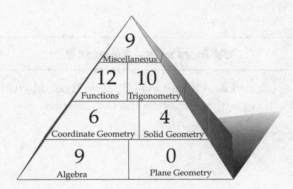

Level IIC

Star System

There's a lot more math you need to know for Level IIC than for Level IC. Most of what's in this book relates to both tests. Any questions, facts, formulas, or strategies that relate to *Level IIC only* are marked with a star.

As you can see, the emphasis in Level IC is on the foundations, while in Level IIC it's more on the advanced topics. But because the more advanced topics are built upon the basics, it can be said that for Level IIC you need to know everything that's tested on Level IC, plus a lot more.

Level of Difficulty and Scoring

The second and third factors to consider in deciding which test to take are *level of difficulty* and scoring. Level IIC questions are considerably more difficult than Level IC questions. Some Level IIC questions are more difficult because they test more advanced topics. But even the Level IIC questions on basic math are generally more difficult than their counterparts on Level IC. This big difference in level of difficulty, however, is partially offset by differences in the score conversion tables. On Level IC, you would probably need to answer every question correctly to get an 800. On Level IIC, however, you can get six or seven questions wrong and still get an 800. On

Top Gun

On Level IIC you can leave several questions unanswered, or even get them wrong, and still get an 800.

Don't Jump!

If you have the background to take Level IIC, don't jump to the conclusion that you'll get a higher score by taking Level IC instead.

Level IC you would need a raw score of more than 20 (out of 50) to get a 500, but on Level IIC you can get a 500 with a raw score as low as about 11.

You don't need so many right answers to achieve a particular score on Level IIC, so don't assume that you'll get a higher score by taking Level IC. If you've had a year of trigonometry and/or precalculus, you might actually find it easier to reach a particular score goal by taking Level IIC.

Reputation

The final factor to consider is *reputation.* Admissions people know how much more math you have to know to get a good score on Level IIC than on Level IC. The purpose of the SAT II: Subject Tests is to demonstrate how much you've learned in high school. If you've learned enough math to take Level IIC, then show it off!

What's Next

Whether you're taking Level IC or Level IIC, you'll need to use a calculator strategically. We'll cover that subject in the next chapter.

CHAPTER 3

The Calculator

In this chapter we'll take an important look at the calculator. We'll discuss what kind to use and how and when to use it during the SAT II: Mathematics Subject Tests.

What to Do Before the Test

First of all, get a scientific calculator. Don't try to take a Math Subject Test without a calculator. That's what the "C" in Level IC and Level IIC stands for. It's not like the SAT I, for which a calculator is permitted, but not really necessary. On the SAT: II Mathematics Subject Tests, a calculator is essential.

Know what kind of calculator you should use. Almost any kind of calculator is allowed, even programmable and graphing calculators. Excluded are laptops and other minicomputers, or machines that print, make noise, or need to be plugged in. Otherwise almost anything goes. Just be sure your calculator performs the following:

- Sine, cosine, and tangent
- Arcsine, arccosine, and arctangent
- Squares, cubes, and other powers
- Square roots and cube roots
- Base-10 logarithms

If you're taking Level IIC, you'll also want your calculator to perform these functions:

- Radians
- Natural logarithms

Not Allowed

Don't even try using laptops and other minicomputers, or machines that print, make noise, or need to be plugged in.

Not

Just as important as knowing when and how to use your calculator is knowing when and how *not* to use your calculator.

Get to know your calculator. Getting the right calculator is only the first step. You must then get used to working with it. Going into the test with an unfamiliar calculator is not much better than going in with no calculator at all.

Practice using your calculator on testlike questions. It's not enough to know how to use your calculator. You need to know how to use it effectively on SAT II: Math questions. Whenever you work with this book, whenever you're taking a practice test, practice with the very calculator you will use on Test Day. With some experience you will learn when to use and when *not* to use, how to use and how *not* to use your calculator on test questions.

Make sure your calculator is in good working order. You'll feel a lot more confident if you put new batteries in your calculator the night before the test and then check it out to see that it's working properly. You should also take spare batteries with you to the test.

Using Your Calculator During the Test

Here are some tips for using your calculator strategically during the test.

Don't use your calculator too often.

You will not need your calculator for every question. Top scorers use their calculators on fewer than half the questions. This is still a math test, not a calculator test. Success depends more on your problem-solving skills than on your calculator skills. The calculator is just one tool. Use it wisely and sparingly.

According to the test makers, your calculator will be of little or no use on about 60 percent of the questions on the test. Here's an example where the calculator is no help:

1. If $x \neq \pm 1$, then $\dfrac{x+1}{x-1} - \dfrac{x-1}{x+1} =$

    (A) $\dfrac{2x}{x-1}$ (B) $\dfrac{2x}{x^2+1}$ (C) $\dfrac{2x}{x^2-1}$ (D) $\dfrac{4x}{x^2+1}$ (E) $\dfrac{4x}{x^2-1}$

To answer this question you need to be adept at algebraic manipulation. Your calculator won't help here. To subtract fractions, even algebraic fractions like these, you need a common denominator, which in this case is the product of the denominators $x - 1$ and $x + 1$:

This Is Not a Brain

Your calculator can't do your thinking for you.

$$\frac{x+1}{x-1} - \frac{x-1}{x+1} = \frac{(x+1)(x+1)}{(x-1)(x+1)} - \frac{(x-1)(x-1)}{(x-1)(x+1)}$$

$$= \frac{x^2 + 2x + 1}{x^2 - 1} - \frac{x^2 - 2x + 1}{x^2 - 1}$$

$$= \frac{x^2 + 2x + 1 - x^2 + 2x - 1}{x^2 - 1}$$

$$= \frac{4x}{x^2 - 1}$$

The answer is (E).

Here's a question for which you have a choice. You can answer it just as quickly and easily with or without a calculator:

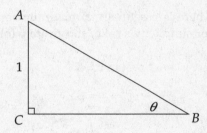

Figure 1

2.   In Figure 1, if $\sin \theta = 0.5$, what is the length of BC?

(A)   1.25

(B)   1.41

(C)   1.50

(D)   1.73

(E)   2.00

You could use your calculator to find out what angle has a sine of 0.5, but you might just know that it's 30°. This is a 30-60-90 triangle, and BC is the longer leg, which is equal to the shorter leg times $\sqrt{3}$. Since $AC = 1$, then $BC = \sqrt{3}$. You could use your calculator to find the square root of 3, but you should probably know that it's about 1.73. The answer is (D).

Don't Get Punchy

Don't start punching calculator keys until you've given the problem a little thought.

Don't use your calculator too early.

When you do use your calculator, it will usually be at a later stage in solving a problem. Never start punching calculator keys before you've given the problem a little thought. Know what you're doing, and where you're going, before you start calculating.

Here's a question that really requires a calculator, but not until a late problem-solving stage:

3.   If $0° < x < 90°$, and $\tan^2 x - \tan x = 6$, which of the following could be the value of x in degrees?

(A)   63

(B)   67

(C)   72

(D)   77

(E)   81

What you have here is essentially a quadratic equation in which the unknown is $\tan x$. For simplicity's sake, substitute y for $\tan x$:

$$\tan^2 x - \tan x = 6$$

$$y^2 - y = 6$$

Now solve for y:

$$y^2 - y = 6$$

$$y^2 - y - 6 = 0$$

$$(y - 3)(y + 2) = 0$$

$$y = 3 \text{ or } -2$$

Now you know that $\tan x = 3$ or -2. Since x is a positive acute angle, the tangent is positive, and $\tan x = 3$. Now's the time to use your calculator. You want to know what acute angle has a tangent of 3. In other words, what you're looking for is the arctangent of 3. Nobody expects you to know the value of arctan(3) off the top of your head. Use your calculator:

$$\tan x = 3$$

$$x = \arctan(3) \approx 71.57$$

The answer is (C).

Pay attention as you read further to how the calculator should be used in answering SAT II: Math questions. Keep your trusty calculator at your side as you proceed, but remember to practice restraint.

What's Next

Now you've covered the basics about what's on the tests and how to wield a calculator. In the next section, we'll plunge into our review of the math that's tested. We'll start with algebra in the next chapter.

Facts, Formulas, and Strategies

This section reviews the content that's tested on Level IC and Level IIC of the SAT II: Mathematics Subject Test. The chapters include Diagnostic Tests to help you identify the topics you need to study, and Follow-Up Tests so that you can track your progress.

CLOSE TO HOME   JOHN McPHERSON

A diabolical new testing technique:
math essay questions.

CHAPTER 4

Algebra

No matter which Mathematics Subject Test you're taking, you need what's in this chapter. Level IC and Level IIC both include a lot of algebra. According to the test makers' official breakdown, 30 percent of Level IC questions and 18 percent of Level IIC questions are algebra questions. But that's counting only the ones that are explicitly and primarily algebra questions. In fact, almost all questions on both levels involve algebra. Most functions questions and coordinate geometry questions are algebraic. Most plane geometry and solid geometry questions involve algebraic formulas, and many use algebraic expressions for lengths and angle measures. Most word problems are best solved algebraically. *Algebra is fundamental to SAT II: Mathematics. You can't get a good score without it.*

How to Use This Chapter

Maybe you know all the algebra you need. Find out by taking the Algebra Diagnostic Test beginning on p. 38. The ten Diagnostic questions are typical of the Mathematics Subject Tests. Check your answers using the answer key following the test. No matter how you score, don't worry! The answer key also shows where to find a detailed explanation for each question. The "Find Your Study Plan" section that follows the test will suggest next steps based on your performance on the Diagnostic.

 Find Your Level

How you use this chapter really depends on which test you're taking and how much time you have to prep. Find your level and pace below.

Taking the Mathematics Level IC Test? No matter how well you do on the Algebra Diagnostic Test, go on and read everything in this chapter and do all the practice problems.

 Algebra Facts and Formulas in This Chapter

- The Five Rules of Exponents (p. 44)
- Combining Like Terms (p. 45)
- Multiplying Monomials (p. 45)
- Multiplying Binomials— FOIL (p. 45)
- Classic Factorables (p. 48)
- Quadratic Formula (p. 53)
- Inequalities and Absolute Value (p. 56)

 What's This Mean?

This icon flags sections that tell you how to use the chapter to fit your study plan.

Kaplan Strategies in This Chapter

- Pick numbers (p. 49).
- Backsolve (p. 52).
- Don't do more work than you have to (p. 55).

Mathematics Level IC Shortcut If you can answer at least seven of the ten questions on the Algebra Diagnostic Test correctly, then you already know the material in this chapter well enough to move on.

Taking the Mathematics Level IIC Test? No matter how well you do on the Diagnostic Test, go on and read everything in this chapter and do all the practice problems.

Mathematics Level IIC Shortcut If you can answer at least seven of the ten questions on the Algebra Diagnostic Test correctly, then you already know the material in this chapter well enough to move on.

Panic Plan? The material in this chapter is vital. If you can't get most of the questions on the Algebra Diagnostic Test right, you should study this chapter—even if it's just two days before Test Day.

Algebra Diagnostic Test

10 Questions (12 Minutes)

Directions: Solve the following problems and choose the best answer from those given. Fill in the oval corresponding to the best answer choice in the grid to the right of each question. (Answers are on page 42.)

DO YOUR FIGURING HERE.

1.   If $x = y^2$ and $y = \dfrac{5}{k}$, what is the value of x when $k = \dfrac{1}{2}$?

   (A)   2.50

   (B)   3.16

   (C)   6.25

   (D)   20.00

   (E)   100.00

   Ⓐ Ⓑ Ⓒ Ⓓ Ⓔ

$x = \left(\dfrac{5}{k}\right)^2$

$x = \dfrac{25}{k^2}$

$x = \dfrac{25}{\frac{1}{4}}$

$x = \dfrac{25}{1} \times \dfrac{4}{1}$

$\dfrac{225}{4}$

$\overline{}$

DO YOUR FIGURING HERE.

2. For all $xyz \neq 0$, $\dfrac{6x^2 y^{12} z^6}{\left(2x^2 yz\right)^3} =$

(A) $\dfrac{y^4 z^2}{x^3}$

(B) $\dfrac{y^9 z^4}{x^4}$

(C) $\dfrac{y^9 z^3}{2x^3}$

(D) $\dfrac{3y^4 z^2}{4x^3}$

(E) $\dfrac{3y^9 z^3}{4x^4}$

ⒶⒷⒸⒹ**Ⓔ**

3. When $2x^3 + 3x^2 - 4x + k$ is divided by $x + 2$, the remainder is 3. What is the value of k ?

(A) –1   (B) 1   (C) 2   (D) 3   (E) 5

ⒶⒷⒸⒹⒺ

4. For all $x \neq \pm 3$, $\dfrac{3x^2 - 11x + 6}{9 - x^2} =$

(A) $\dfrac{2 - 3x}{x - 3}$

(B) $\dfrac{2 - 3x}{x + 3}$

(C) $\dfrac{2x - 3}{x - 3}$

(D) $\dfrac{3x - 2}{x + 3}$

(E) $\dfrac{3x - 2}{x - 3}$

ⒶⒷⒸⒹⒺ

39

DO YOUR FIGURING HERE.

5. If $\sqrt[3]{8x+6} = -3$, what is the value of x?

(A) −4.125

(B) −2.625

(C) −1.875

(D) −1.125

(E) 2.625

Ⓐ Ⓑ Ⓒ Ⓓ Ⓔ

6. If $\dfrac{5}{x+3} = \dfrac{1}{x} + \dfrac{1}{2x}$, what is the value of x?

(A) $\dfrac{3}{14}$ (B) $\dfrac{1}{3}$ (C) $\dfrac{6}{13}$ (D) $\dfrac{3}{4}$ (E) $\dfrac{9}{7}$

Ⓐ Ⓑ Ⓒ Ⓓ Ⓔ

7. If $8^x = 16^{x-1}$, then $x =$

(A) $\dfrac{1}{8}$ (B) $\dfrac{1}{2}$ (C) 2   (D) 4   (E) 8

Ⓐ Ⓑ Ⓒ Ⓓ Ⓔ

8. If $a = \dfrac{b+x}{c+x}$, what is the value of x in terms of a, b, and c?

(A) $\dfrac{a-bc}{a-1}$

(B) $\dfrac{b-ac}{a-1}$

(C) $\dfrac{a+bc}{a+1}$

(D) $\dfrac{ac+b}{a+1}$

(E) $\dfrac{ac-b}{a}$

Ⓐ Ⓑ Ⓒ Ⓓ Ⓔ

DO YOUR FIGURING HERE.

9. If $2x - 9y = 11$ and $x + 12y = -8$, what is the value of $x + y$?

    (A) $-\dfrac{29}{11}$

    (B) $-\dfrac{9}{11}$

    (C) 1

    (D) $\dfrac{20}{11}$

    (E) $\dfrac{29}{11}$ Ⓐ Ⓑ Ⓒ Ⓓ Ⓔ

10. Which of the following is the solution set of $|2x - 3| < 7$?

    (A) $\{x: -5 < x < 2\}$

    (B) $\{x: -5 < x < 5\}$

    (C) $\{x: -2 < x < 5\}$

    (D) $\{x: x < -5 \text{ or } x > 2\}$

    (E) $\{x: x < -2 \text{ or } x > 5\}$ Ⓐ Ⓑ Ⓒ Ⓓ Ⓔ

STOP! **END OF TEST. DO NOT TURN THE PAGE UNTIL YOU ARE READY TO CHECK YOUR ANSWERS.**

Diagnostic Test Answers and Reviews

1. E
See "Evaluating Expressions," p. 42.
2. E
See "Exponents—Key Operations," p. 43.
3. A
See "Dividing Polynomials," p. 46.
4. B
See "Factoring," p. 47.
5. A
See "The Golden Rule of Equations," p. 49.
6. E
See "Unknown in a Denominator," p. 50.
7. D
See "Unknown in an Exponent," p. 52.
8. B
See "In Terms Of," p. 54.
9. C
See "Simultaneous Equations," p. 55.
10. C
See "Inequalities and Absolute Value," p. 56.

Test Topics

This icon appears next to each discussion of a math topic that's tested on the SAT II.

Find Your Study Plan

The answer key shows where in this chapter to find explanations for the questions you missed. Here's how you should proceed based on your Diagnostic Test score.

9–10: Superb! You really know your algebra. Unless you have lots of time and just love to read about algebra, you might consider skipping this chapter. You seem to know it all already. To make absolutely sure, you could look over the facts, formulas, and strategies in the margins of this chapter. And if you just want to do some more algebra problems, go to the Follow-Up Test at the end of this chapter.

7–8: Excellent! You're quite good at algebra. Some of these are especially difficult questions. If you're taking a "shortcut" or you're on the Panic Plan, you don't really have time to study this chapter, and you don't really need to. You might want to look at those pages that address the questions you didn't get right. And if you just want to do some more algebra problems, go to the Follow-Up Test at the end of the chapter.

0–6: No matter which level test you're taking, or how pressed for time you are, you should continue to read this chapter and do the Follow-Up Test at the end. You need to brush up on your algebra before you can take full advantage of later chapters.

Algebra Test Topics

The questions in the Algebra Diagnostic Test are typical of the Mathematics Subject Tests. They cover a wide range of algebra topics, from simple expressions to quadratic equations to inequalities with absolute value signs. In this chapter we'll use these questions to review the algebra you need to know for the Mathematics Subject Tests. We will also use these questions to demonstrate effective problem-solving techniques, alternative methods, and test-taking strategies that apply to SAT II algebra questions.

Evaluating Expressions

The simplest and most straightforward type of algebra questions you will encounter on the SAT II are like Example 1:

Example 1

If $x = y^2$ and $y = \dfrac{5}{k}$, what is the value of x when $k = \dfrac{1}{2}$?

(A)   2.50

(B)   3.16

(C)   6.25

(D) 20.00

(E) 100.00

Answering a question like this is basically just a matter of plugging in and cranking out. You are given an expression for x in terms of y, an expression for y in terms of k, and the value of k. Plug $k = \dfrac{1}{2}$ into the second equation to find y :

$$y = \frac{5}{k} = \frac{5}{\frac{1}{2}} = 5 \div \frac{1}{2} = \frac{5}{1} \times \frac{2}{1} = 10$$

Now plug $y = 10$ into the first equation to find x :

$$x = y^2 = 10^2 = 100$$

The answer is (E).

 ### Exponents—Key Operations

You can't be adept at algebra unless you're completely at ease with exponents. Here's what you need to know:

Multiplying powers with the same base: To multiply powers with the same base, keep the base and add the exponents:

$$x^3 \times x^4 = x^{3+4} = x^7$$

Dividing powers with the same base: To divide powers with the same base, keep the base and subtract the exponents:

$$y^{13} \div y^8 = y^{13-8} = y^5$$

Raising a power to an exponent: To raise a power to an exponent, keep the base and multiply the exponents:

$$(x^3)^4 = x^{3 \times 4} = x^{12}$$

> **"You can't be adept at algebra unless you're completely at ease with exponents."**

The Five Rules of Exponents

1. $(x^m)(x^n) = x^{m+n}$

2. $\dfrac{x^m}{x^n} = x^{m-n}$

3. $(x^m)^n = x^{mn}$

4. $(x^n)(y^n) = (xy)^n$

5. $\dfrac{x^n}{y^n} = \left(\dfrac{x}{y}\right)^n$

Multiplying powers with the same exponent: To multiply powers with the same exponent, multiply the bases and keep the exponent:

$$(3^x)(4^x) = 12^x$$

Dividing powers with the same exponent: To divide powers with the same exponent, divide the bases and keep the exponent:

$$\frac{6^x}{2^x} = 3^x$$

On Test Day, you might encounter an algebra question, like Example 2, that specifically tests your understanding of the rules of exponents.

Example 2

For all $xyz \neq 0$, $\dfrac{6x^2 y^{12} z^6}{\left(2x^2 yz\right)^3} =$

(A) $\dfrac{y^4 z^2}{x^3}$ (B) $\dfrac{y^9 z^4}{x^4}$ (C) $\dfrac{y^9 z^3}{2x^3}$ (D) $\dfrac{3y^4 z^2}{4x^3}$ (E) $\dfrac{3y^9 z^3}{4x^4}$

There's nothing tricky about this question if you know how to work with exponents. The first step is to eliminate the parentheses. Everything inside gets cubed:

$$\frac{6x^2 y^{12} z^6}{\left(2x^2 yz\right)^3} = \frac{6x^2 y^{12} z^6}{8x^6 y^3 z^3}$$

The next step is to look for factors common to the numerator and denominator. The 6 on top and the 8 on bottom reduce to 3 over 4—so it already looks like the answer's going to be (D) or (E). The x^2 on top cancels with the x^6 on bottom, leaving x^4 on bottom. (You're actually subtracting the exponents: $2 - 6 = -4$, since x^{-4} is the same as $\dfrac{1}{x^4}$.) The y^{12} on top cancels with the y^3 on bottom, leaving y^9 on top. And the z^6 on top cancels with the z^3 on bottom, leaving z^3 on top:

$$\frac{6x^2 y^{12} z^6}{8x^6 y^3 z^3} = \frac{3y^9 z^3}{4x^4}$$

The answer is (E).

Adding, Subtracting, and Multiplying Polynomials

Algebra is the basic language of the Mathematics Subject Tests, and you will want to be fluent in that language. You might not get a whole lot of questions that ask explicitly about such basic algebra procedures as combining like terms, multiplying binomials, or factoring algebraic expressions, but you will do all of those things in the course of working out the answers to more advanced questions. So it's essential that you be at ease with the mechanics of algebraic manipulations.

Combining like terms: To combine like terms, keep the variable part unchanged while adding or subtracting the coefficients:

$$2a + 3a = (2 + 3)a = 5a$$

Adding or subtracting polynomials: To add or subtract polynomials, combine like terms:

$$(3x^2 + 5x - 7) - (x^2 + 12) =$$
$$(3x^2 - x^2) + 5x + (-7 - 12) =$$
$$2x^2 + 5x - 19$$

Multiplying monomials: To multiply monomials, multiply the coefficients and the variables separately:

$$2x \times 3x = (2 \times 3)(x \times x) = 6x^2$$

Multiplying binomials: To multiply binomials, use FOIL. To multiply $(x + 3)$ by $(x + 4)$, first multiply the **F**irst terms: $x \times x = x^2$. Next the **O**uter terms: $x \times 4 = 4x$. Then the **I**nner terms: $3 \times x = 3x$. And finally the **L**ast terms: $3 \times 4 = 12$. Then add and combine like terms:

$$x^2 + 4x + 3x + 12 = x^2 + 7x + 12$$

Multiplying polynomials: To multiply polynomials with more than two terms, make sure you multiply each term in the first polynomial by each term in the second. (FOIL works only when you want to multiply two binomials.)

$$\left(x^2 + 3x + 4\right)(x + 5) = x^2(x + 5) + 3x(x + 5) + 4(x + 5)$$
$$= x^3 + 5x^2 + 3x^2 + 15x + 4x + 20$$
$$= x^3 + 8x^2 + 19x + 20$$

After multiplying two polynomials together, the number of terms in your expression before simplifying should equal the number of terms in one polynomial multiplied by the number of terms in the second. In the

Combining Like Terms

$$ax + bx = (a + b)x$$

$$ax - bx = (a - b)x$$

Multiplying Monomials

$$(ax)(bx) = (ab)x^2$$

Multiplying Binomials—FOIL

$(a + b)(c + d) = ?$
First = ac
    Outer = ad
        Inner = bc
            Last = bd
Product = $ac + ad + bc + bd$

example above, you should have $3 \times 2 = 6$ terms in the product before you simplify like terms.

Dividing Polynomials

To divide polynomials, you can use long division. For example, to divide $2x^3 + 13x^2 + 11x - 16$ by $x + 5$:

$$x + 5 \overline{)2x^3 + 13x^2 + 11x - 16}$$

The first term of the quotient is $2x^2$, because that's what will give you a $2x^3$ as a first term when you multiply it by $x + 5$:

$$
\begin{array}{r}
2x^2 \\
x + 5 \overline{)2x^3 + 13x^2 + 11x - 16} \\
2x^3 + 10x^2
\end{array}
$$

Subtract and continue in the same way as when dividing numbers:

$$
\begin{array}{r}
2x^2 + 3x - 4 \\
x + 5 \overline{)2x^3 + 13x^2 + 11x - 16} \\
2x^3 + 10x^2 \\
\hline
3x^2 + 11x \\
3x^2 + 15x \\
\hline
-4x - 16 \\
-4x - 20 \\
\hline
4
\end{array}
$$

The result is $2x^2 + 3x - 4$ with a remainder of 4.

> **"** Once you know the rules, adding, subtracting, multiplying, and even dividing polynomials is automatic. **"**

Long division is the way to do Example 3:

Example 3

When $2x^3 + 3x^2 - 4x + k$ is divided by $x + 2$, the remainder is 3. What is the value of k ?

(A) –1   (B) 1   (C) 2   (D) 3   (E) 5

To answer this question, start by cranking out the long division:

$$\begin{array}{r} 2x^2 - x - 2 \\ x + 2 \overline{) 2x^3 + 3x^2 - 4x + k} \\ \underline{2x^3 + 4x^2} \\ -x^2 - 4x \\ \underline{-x^2 - 2x} \\ -2x + k \\ \underline{-2x - 4} \end{array}$$

The question says that the remainder is 3, so, whatever k is, when you subtract -4 from it, you get 3:

$$k - (-4) = 3$$
$$k + 4 = 3$$
$$k = -1$$

The answer is (A).

 ## Factoring

Performing operations on polynomials is largely a matter of cranking it out. Once you know the rules, adding, subtracting, multiplying, and even dividing is automatic. Factoring algebraic expressions is a different matter. To factor successfully you have to do more thinking and less cranking. You have to try to figure out what expressions multiplied will give you the polynomial you're looking at. Sometimes that means having a good eye for the test makers' favorite factorables:

- Factor common to all terms
- Difference of squares
- Square of a binomial

Factor common to all terms: A factor common to all the terms of a polynomial can be factored out. This is essentially the distributive property in reverse. For example, all three terms in the polynomial $3x^3 + 12x^2 - 6x$ contain a factor of $3x$. Pulling out the common factor yields $3x(x^2 + 4x - 2)$.

Difference of squares: You will want to be especially keen at spotting polynomials in the form of the difference of squares. Whenever you have two identifiable squares with a minus sign between them, you can factor the expression like this:

$$a^2 - b^2 = (a + b)(a - b)$$

$4x^2 - 9$, for example, factors to $(2x + 3)(2x - 3)$.

> "Factoring requires you to do more thinking and less cranking."

Classic Factorables

Factor common to all terms:

$$ax + ay = a(x + y)$$

Difference of squares:

$$a^2 - b^2 = (a - b)(a + b)$$

Square of binomial:

$$a^2 + 2ab + b^2 = (a + b)^2$$

Squares of binomials: Learn to recognize polynomials that are squares of binomials:

$$a^2 + 2ab + b^2 = (a + b)^2$$

$$a^2 - 2ab + b^2 = (a - b)^2$$

For example, $4x^2 + 12x + 9$ factors to $(2x + 3)^2$, and $a^2 - 10a + 25$ factors to $(a - 5)^2$.

Sometimes you'll want to factor a polynomial that's not in any of these classic factorable forms. When that happens, factoring becomes a kind of logic exercise, with some trial and error thrown in. To factor a quadratic expression, think about what binomials you could use FOIL on to get that quadratic expression. For example, to factor $x^2 - 5x + 6$, think about what First terms will produce x^2, what Last terms will produce $+6$, and what Outer and Inner terms will produce $-5x$. Some common sense—and a little trial and error—will lead you to $(x - 2)(x - 3)$.

Example 4 is a good instance of an SAT II: Mathematics question that calls for factoring.

Example 4

For all $x \neq \pm 3$, $\dfrac{3x^2 - 11x + 6}{9 - x^2} =$

(A) $\dfrac{2 - 3x}{x - 3}$ (B) $\dfrac{2 - 3x}{x + 3}$ (C) $\dfrac{2x - 3}{x - 3}$ (D) $\dfrac{3x - 2}{x + 3}$ (E) $\dfrac{3x - 2}{x - 3}$

To reduce a fraction, you eliminate factors common to the top and bottom. So the first step in reducing an algebraic fraction is to *factor the numerator and denominator*. Here the denominator is easy since it's the difference of squares: $9 - x^2 = (3 - x)(3 + x)$. The numerator takes some thought and some trial and error. For the first term to be $3x^2$, the first terms of the factors must be $3x$ and x. For the last term to be $+6$, the last terms must be either $+2$ and $+3$, or -2 and -3, or $+1$ and $+6$, or -1 and -6. After a few tries, you should come up with: $3x^2 - 11x + 6 = (3x - 2)(x - 3)$. Now the fraction looks like this:

$$\frac{3x^2 - 11x + 6}{9 - x^2} = \frac{(3x - 2)(x - 3)}{(3 - x)(3 + x)}$$

In this form there are no precisely common factors, but there is a factor in the numerator that's the opposite (negative) of a factor in the denominator: $x - 3$ and $3 - x$ are opposites. Factor -1 out of the numerator and get:

$$\frac{(3x-2)(x-3)}{(3-x)(3+x)} = \frac{(-1)(3x-2)(3-x)}{(3-x)(3+x)}$$

Now $(3-x)$ can be eliminated from both the top and the bottom:

$$\frac{(-1)(3x-2)(3-x)}{(3-x)(3+x)} = \frac{-(3x-2)}{3+x} = \frac{-3x+2}{3+x}$$

That's the same as choice (B):

$$\frac{-3x+2}{3+x} = \frac{2-3x}{x+3}$$

Alternative method: Here's another way to answer this question. *Pick a number for* x *and see what happens.* One of the answer choices will give you the same value as the original fraction will, no matter what you plug in for x. Pick a number that's easy to work with—like 0. When you plug $x = 0$ into the original expression, any term with an x drops out, and you end up with $\frac{6}{9}$, or $\frac{2}{3}$. Now plug $x = 0$ into each answer choice to see which ones equal $\frac{2}{3}$. When you get to (B), it works, but you can't stop there. It might just be a coincidence. When you pick numbers, *look at every answer choice*. Choice (E) also works for $x = 0$. At least you know one of those is the correct answer, and you can decide between them by picking another value for x.

This is not a sophisticated approach, but who cares? You don't get points for elegance. You get points for right answers.

 ## The Golden Rule of Equations

You probably remember the basic procedure for solving algebraic equations: *Do the same thing to both sides.* You can do almost anything you want to one side of an equation as long as you preserve the equality by doing the same thing to the other side. Your aim in whatever you do to both sides is to get the variable (or expression) you're solving for all by itself on one side. Look at Example 5:

 Pick Numbers.

When the answer choices are algebraic expressions, it often works to *pick numbers* for the unknowns, plug those numbers into the stem and see what you get, and then plug those same numbers into the answer choices to find matches.

Warning: When you pick numbers, you have to check *all* the answer choices. Sometimes more than one works with the number(s) you pick, in which case you have to pick numbers again.

> "Solving linear equations is usually pretty straight-forward."

Example 5

If $\sqrt[3]{8x+6} = -3$, what is the value of x?

(A) −4.125   (B) −2.625   (C) −1.875   (D) −1.125   (E) 2.625

To solve this equation for x means to do whatever you have to to both sides of the equation to get x all by itself on one side. Layer by layer you want to peel away all those extra symbols and numbers around the x. First you want to get rid of that cube-root symbol. The way to undo a cube root is to cube both sides:

$$\sqrt[3]{8x+6} = -3$$
$$\left(\sqrt[3]{8x+6}\right)^3 = (-3)^3$$
$$8x+6 = -27$$

The rest is easy. Subtract 6 from both sides and divide both sides by 8:

$$8x+6 = -27$$
$$8x = -27-6$$
$$8x = -33$$
$$x = -\frac{33}{8} = -4.125$$

The answer is (A).

The test makers have a couple of favorite equation types that you should be prepared to solve. Solving linear equations is usually pretty straightforward. Generally it's obvious what to do to isolate the unknown. But when the unknown is in a denominator or an exponent, it might not be so obvious how to proceed.

Unknown in a Denominator

The basic procedure for solving an equation is the same even when the unknown is in a denominator: Do the same thing to both sides. In this case you *multiply in order to undo division*. If, for example, you wanted to solve the equation $1+\dfrac{1}{x} = 2-\dfrac{1}{x}$, you would multiply both sides by x:

$$1+\frac{1}{x} = 2-\frac{1}{x}$$
$$x\left(1+\frac{1}{x}\right) = x\left(2-\frac{1}{x}\right)$$
$$x+1 = 2x-1$$

Now you have an equation with no denominators, which is easy to solve:

$$x + 1 = 2x - 1$$
$$x - 2x = -1 - 1$$
$$-x = -2$$
$$x = 2$$

Another good way to solve an equation with the unknown in the denominator is to *cross multiply*. That's the best way to do Example 6.

Example 6

If $\dfrac{5}{x+3} = \dfrac{1}{x} + \dfrac{1}{2x}$, what is the value of x ?

(A) $\dfrac{3}{14}$ (B) $\dfrac{1}{3}$ (C) $\dfrac{6}{13}$ (D) $\dfrac{3}{4}$ (E) $\dfrac{9}{7}$

Before you can cross multiply, you need to reexpress the right side of the equation as a single fraction. That means giving the two fractions a common denominator and adding them. The common denominator is $2x$:

$$\frac{5}{x+3} = \frac{1}{x} + \frac{1}{2x}$$

$$\frac{5}{x+3} = \frac{2}{2x} + \frac{1}{2x}$$

$$\frac{5}{x+3} = \frac{3}{2x}$$

Now you can cross multiply:

$$\frac{5}{x+3} = \frac{3}{2x}$$

$$(5)(2x) = (x+3)(3)$$
$$10x = 3x + 9$$
$$10x - 3x = 9$$
$$7x = 9$$
$$x = \frac{9}{7}$$

The answer is (E).

Backsolve.

Sometimes, when you can't figure out how to attack a problem directly, you may be able to go at it backwards. One of the given answer choices has to be the correct answer, so you can sometimes just try them out. Since the answer choices are generally in numerical order, start with choice (C). If (C) doesn't work, you might at least know whether you're looking for something bigger or something smaller than (C). *Note:* You don't have to try all the answer choices when backsolving. As soon as you find a choice that works, you're finished.

 ## Unknown in an Exponent

The procedure for solving an equation when the unknown is in an exponent is a little different. What you want to do in this situation is to reexpress one or both sides of the equation so that the two sides have the same base. Look at Example 7:

Example 7

If $8^x = 16^{x-1}$, then $x =$

(A) $\dfrac{1}{8}$

(B) $\dfrac{1}{2}$

(C) 2

(D) 4

(E) 8

In this case, the base on the left is 8 and the base on the right is 16. They're both powers of 2, so you can reexpress both sides as powers of 2:

$$8^x = 16^{x-1}$$
$$(2^3)^x = (2^4)^{x-1}$$
$$2^{3x} = 2^{4x-4}$$

Now that both sides have the same base, you can simply set the exponent expressions equal and solve for x :

$$3x = 4x - 4$$
$$3x - 4x = -4$$
$$-x = -4$$
$$x = 4$$

Alternative method: Here's another way to answer this question. Nobody says you have to figure out the answer to the question and then look for your solution among the answer choices. If you don't see how to do it the front way, *try working backwards*. Try plugging the answer choices back into the problem until you find the one that works. Here, if you start with (C) and $x = 2$, you get $8^x = 8^2 = 64$ on the left side of the equation and $16^{x-1} = 16^1 = 16$ on the right side. It's not clear whether (C) was too small or too large, so you should probably try (D) next—it's easier to work with than (B), which is a fraction. If $x = 4$, then $8^x = 8^4 = 4{,}096$ on the left, and $16^{x-1} = 16^3 = 4{,}096$. No need to do any more. (D) works, so it's the answer.

Don't depend on backsolving too much. There are lots of math questions that can't be backsolved at all. And most that *can* be backsolved are almost certainly more *quickly* solved by a more direct approach.

 ## Quadratic Equations

To solve a quadratic equation, put it in the "$ax^2 + bx + c = 0$" form, factor the left side (if you can), and set each factor equal to 0 separately to get the two solutions. To solve $x^2 + 12 = 7x$, first rewrite it as $x^2 - 7x + 12 = 0$. Then factor the left side:

$$x^2 - 7x + 12 = 0$$
$$(x - 3)(x - 4) = 0$$
$$x - 3 = 0 \quad \text{or} \quad x - 4 = 0$$
$$x = 3 \quad \text{or} \quad 4$$

Sometimes the left side may not be obviously factorable. You can always use the *quadratic formula*. Just plug in the coefficients a, b, and c from $ax^2 + bx + c = 0$ into the formula:

$$x = \frac{-b \pm \sqrt{b^2 - 4ac}}{2a}$$

To solve $x^2 + 4x + 2 = 0$, plug $a = 1$, $b = 4$, and $c = 2$ into the formula:

$$x = \frac{-4 \pm \sqrt{4^2 - 4 \cdot 1 \cdot 2}}{2 \cdot 1}$$
$$= \frac{-4 \pm \sqrt{8}}{2} = -2 \pm \sqrt{2}$$

 ### Quadratic Formula

If $ax^2 + bx + c = 0$, then:

$$x = \frac{-b \pm \sqrt{b^2 - 4ac}}{2a}$$

"In Terms Of"

So far in this chapter solving an equation has meant finding a numerical value for the unknown. When there's more than one variable, it's generally impossible to get numerical solutions. Instead, what you do is solve for the unknown *in terms of* the other variables.

To solve an equation for one variable in terms of another means to isolate the one variable on one side of the equation, leaving an expression containing the other variable on the other side of the equation. For example, to solve the equation $3x - 10y = -5x + 6y$ for x in terms of y, isolate x:

$$3x - 10y = -5x + 6y$$
$$3x + 5x = 6y + 10y$$
$$8x = 16y$$
$$x = 2y$$

Now look at the next SAT II example, which asks you to solve "in terms of."

Example 8

If $a = \dfrac{b+x}{c+x}$, what is the value of x in terms of a, b, and c?

(A) $\dfrac{a-bc}{a-1}$ (B) $\dfrac{b-ac}{a-1}$ (C) $\dfrac{a+bc}{a+1}$ (D) $\dfrac{ac+b}{a+1}$ (E) $\dfrac{ac-b}{a}$

You want to get x on one side by itself. First thing to do is eliminate the denominator by multiplying both sides by $c + x$:

$$a = \frac{b+x}{c+x}$$
$$a(c+x) = \left(\frac{b+x}{c+x}\right)(c+x)$$
$$ac + ax = b + x$$

Next move all terms with x to one side and all terms without to the other:

$$ac + ax = b + x$$
$$ax - x = b - ac$$

Now factor x out of the left side and divide both sides by the other factor to isolate x:

$$ax - x = b - ac$$
$$x(a-1) = b - ac$$
$$x = \frac{b-ac}{a-1}$$

The answer is (B).

Simultaneous Equations

You can get numerical solutions for more than one unknown if you are given more than one equation. The test makers like simultaneous equations questions because they take a little thought to answer. Solving simultaneous equations almost always involves combining equations, but you have to figure out what's the best way to combine the equations.

You can solve for two variables only if you have two distinct equations. Two forms of the same equation will not be adequate. Combine the equations in such a way that one of the variables cancels out. For example, to solve the two equations $4x + 3y = 8$ and $x + y = 3$, multiply both sides of the second equation by -3 to get $-3x - 3y = -9$. Now add the two equations; the $3y$ and the $-3y$ cancel out, leaving $x = -1$. Plug that back into either one of the original equations and you'll find that $y = 4$.

Example 9 is a simultaneous equations question:

Example 9

If $2x - 9y = 11$ and $x + 12y = -8$, what is the value of $x + y$?

(A) $-\dfrac{29}{11}$ (B) $-\dfrac{9}{11}$ (C) 1 (D) $\dfrac{20}{11}$ (E) $\dfrac{29}{11}$

If you just plow ahead without thinking, you might try to answer this question by solving for one variable at a time. That would work, but it would take a lot more time than this question needs. As usual, the key to this simultaneous equations question is to combine the equations, but combining the equations doesn't necessarily mean losing a variable. Look what happens here if you just add the equations as presented:

$$2x - 9y = 11$$
$$+[x + 12y = -8]$$
$$\overline{3x + 3y = 3}$$

Suddenly you're almost there! Just divide both sides by 3 and you get $x + y = 1$. The answer is (C).

 ## Absolute Value

To solve an equation that includes absolute value signs, think about the two different cases. For example, to solve the equation $\left| x - 12 \right| = 3$, think of it as two equations:

$$x - 12 = 3 \text{ or } x - 12 = -3$$

$$x = 15 \text{ or } 9$$

Don't Do More Work Than You Have To.

You don't always have to find the value of each variable to answer a simultaneous equations question.

Inequalities

To solve an inequality, do whatever is necessary to both sides to isolate the variable. Just remember that when you multiply or divide both sides by a negative number, you must reverse the sign. To solve $-5x + 7 < -3$, subtract 7 from both sides to get $-5x < -10$. Now divide both sides by -5, remembering to reverse the sign: $x > 2$.

 Inequalities and Absolute Value

About the most complicated algebraic solving you'll have to do on the Math Subject Tests will involve inequalities and absolute value signs. Look at Example 10:

Example 10

Which of the following is the solution set of $|2x - 3| < 7$?

(A) $\{x: -5 < x < 2\}$

(B) $\{x: -5 < x < 5\}$

(C) $\{x: -2 < x < 5\}$

(D) $\{x: x < -5 \text{ or } x > 2\}$

(E) $\{x: x < -2 \text{ or } x > 5\}$

Inequalities and Absolute Value

If $n > 0$,

$$|\text{whatever}| < n$$
$$\Downarrow$$
$$-n < \text{whatever} < n$$

$$|\text{whatever}| > n$$
$$\Downarrow$$
$$\text{whatever} < -n \text{ or whatever} > n$$

What does it mean if $|2x - 3| < 7$? It means that if the expression between the absolute value bars is positive, it's less than +7, or, if the expression between the bars is negative, it's greater than -7. In other words, $2x - 3$ is between -7 and $+7$:

$$-7 < 2x - 3 < 7$$
$$-4 < 2x < 10$$
$$-2 < x < 5$$

The answer is (C).

In fact, there's a general rule that applies here: To solve an inequality in the form $|\text{whatever}| < p$, where $p > 0$, just put that "whatever" inside the range $-p$ to p :

$$|\text{whatever}| < p \text{ means: } -p < \text{whatever} < p$$

For example, $|x - 5| < 14$ becomes $-14 < x - 5 < 14$.

And here's another general rule: To solve an inequality in the form $|\text{whatever}| > p$, where $p > 0$, just put that "whatever" outside the range $-p$ to p:

$$|\text{whatever}| > p \text{ means: whatever} < -p \text{ OR whatever} > p$$

For example, $\left|\dfrac{3x+9}{2}\right| > 7$ becomes $\dfrac{3x+9}{2} < -7$ OR $\dfrac{3x+9}{2} > 7$.

Well, you've seen a lot of algebra in this chapter. You've seen ten of the test makers' favorite algebra situations. You've reviewed all the relevant algebra facts and formulas. And you've learned some effective Kaplan test-taking strategies. Now it's time to take the Algebra Follow-Up Test to find out how much you've learned.

$-7 < 2x - 3 < 7$

$-4 < 2x < 10$

$-2 < x < 5$

Algebra Follow-Up Test

10 Questions (12 Minutes)

Directions: Solve the following problems and choose the best answer from those given. Fill in the oval corresponding to the best answer choice in the grid to the right of each question. (Answers and explanations begin on page 62.)

DO YOUR FIGURING HERE.

1.   If $x = 3 - y^2$ and $y = -2$, what is the value of x ?

(A) –2

(B) –1

(C) 1

(D) 2

(E) 25

Ⓐ Ⓑ Ⓒ Ⓓ Ⓔ

2.   For all x, $2^x + 2^x + 2^x + 2^x =$

(A) 2^{x+2}

(B) 2^{x+4}

(C) 2^{3x}

(D) 2^{4x}

(E) 2^{5x}

Ⓐ Ⓑ Ⓒ Ⓓ Ⓔ

DO YOUR FIGURING HERE.

3. For all $x \neq \pm\frac{1}{2}$, $\dfrac{6x^2 - x - 2}{4x^2 - 1} =$

(A) $\dfrac{2 - 3x}{2x + 1}$

(B) $\dfrac{3x + 2}{2x + 1}$

(C) $\dfrac{3x - 2}{2x + 1}$

(D) $\dfrac{3x + 2}{2x - 1}$

(E) $\dfrac{3x - 2}{2x - 1}$

Ⓐ Ⓑ Ⓒ Ⓓ Ⓔ

4. What is the remainder when $3x^3 - 7x + 7$ is divided by $x + 2$?

(A) –5   (B) –3   (C) 1   (D) 3   (E) 5

Ⓐ Ⓑ Ⓒ Ⓓ Ⓔ

5. If $\sqrt[4]{\dfrac{x+1}{2}} = \dfrac{1}{2}$, then $x =$

(A)   –0.969

(B)   –0.875

(C)   0

(D)   0.875

(E)   0.969

Ⓐ Ⓑ Ⓒ Ⓓ Ⓔ

6. If $\dfrac{19}{5x + 17} = \dfrac{19}{31}$, then $x =$

(A)   0.4

(B)   1.4

(C)   2.8

(D)   3.4

(E)   3.8

Ⓐ Ⓑ Ⓒ Ⓓ Ⓔ

59

DO YOUR FIGURING HERE.

7. If $(3^{x^2})(9^x)(3) = 27$ and $x > 0$, what is the value of x ?

(A) 0.268

(B) 0.414

(C) 0.732

(D) 1.414

(E) 1.464

Ⓐ Ⓑ Ⓒ Ⓓ Ⓔ

$3^{x^2} \cdot 9^x_y \cdot 3$

$81^{x^3} = 27$

8. If $y \neq 4a$, and $x = \dfrac{y+a^2}{y-4a}$, what is the value of y in terms of a and x ?

(A) $\dfrac{4a - a^2 x}{x + 1}$

(B) $\dfrac{a^2 - 4ax}{x + 1}$

(C) $\dfrac{a^2 + 4ax}{x + 1}$

(D) $\dfrac{a^2 + 4ax}{x - 1}$

(E) $\dfrac{a^2 - 4ax}{x - 1}$

Ⓐ Ⓑ Ⓒ Ⓓ Ⓔ

9. If one of the following choices is the solution to the pair of equations $4x + ky = 15$ and $x - ky = -25$, which one is it?

(A) $x = -3$ and $y = -5$

(B) $x = -2$ and $y = 3$

(C) $x = 0$ and $y = -2$

(D) $x = 2$ and $y = 3$

(E) $x = 3$ and $y = 5$

Ⓐ Ⓑ Ⓒ Ⓓ Ⓔ

$4x + ky = 15$

$x - ky = -25$

$x = -25 + ky$

$-3 = -25 - 5k$

$-5k = 22$

$k = \dfrac{-22}{5}$

$-12 + 22 =$

$\dfrac{22}{2}$

$\dfrac{12}{10}$

DO YOUR FIGURING HERE.

10. How many integers are in the solution set of $|4x + 3| < 8$?

   (A) None

   (B) Two

   (C) Three

   (D) Four

   (E) Infinitely many

   Ⓐ Ⓑ Ⓒ Ⓓ Ⓔ

$-8 < 4x + 3 < 8$

$-11 < 4x < 5$

4, 0

STOP! **END OF TEST. DO NOT TURN THE PAGE UNTIL YOU ARE READY TO CHECK YOUR ANSWERS.**

Follow-Up Test—Answers and Explanations

Answer Key   1. *B*   2. *A*   3. *E*   4. *B*   5. *B*   6. *C*   7. *C*   8. *D*   9. *B*   10. *D*

1. **(B)**—Plug $y = -2$ into the first equation:

$$x = 3 - y^2 = 3 - (-2)^2 = 3 - 4 = -1$$

2. **(A)**—The sum of 4 identical quantities is 4 times one of those quantities, so the sum of the four terms 2^x is 4 times 2^x:

$$2^x + 2^x + 2^x + 2^x = 4(2^x) = 2^2(2^x) = 2^{x+2}$$

3. **(E)**—Factor the top and the bottom and cancel the factors they have in common:

$$\frac{6x^2 - x - 2}{4x^2 - 1} = \frac{(3x-2)(2x+1)}{(2x-1)(2x+1)} = \frac{3x-2}{2x-1}$$

4. **(B)**—Use long division. Watch out: The expression that goes under the division sign needs a place-holding $0x^2$ term:

$$
\begin{array}{r}
3x^2 - 6x + 5 \\
x+2\overline{)3x^3 + 0x^2 - 7x + 7} \\
\underline{3x^3 + 6x^2} \\
-6x^2 - 7x \\
\underline{-6x^2 - 12x} \\
5x + 7 \\
\underline{5x + 10} \\
-3
\end{array}
$$

The remainder is –3.

5. **(B)**—To undo the fourth-root symbol, raise both sides to the fourth power:

$$\sqrt[4]{\frac{x+1}{2}} = \frac{1}{2}$$

$$\left(\sqrt[4]{\frac{x+1}{2}}\right)^4 = \left(\frac{1}{2}\right)^4$$

$$\frac{x+1}{2} = \frac{1}{16}$$

Now cross multiply:

$$\frac{x+1}{2} = \frac{1}{16}$$
$$(x+1)(16) = (2)(1)$$
$$16x + 16 = 2$$
$$16x = -14$$
$$x = -\frac{14}{16} = -\frac{7}{8} = -0.875$$

6. **(C)**—Don't do more work than you have to. This might look at first glance like a candidate for cross multiplication, but that would just make things more complicated than they need to be. Notice that the fractions on both sides have the same numerator, 19. So the numerator is irrelevant. If the two fractions are equal and they have the same numerator, then they must have the same denominator, so just write an equation that says that one denominator is equal to the other denominator:

$$\frac{19}{5x+17} = \frac{19}{31}$$

$$5x + 17 = 31$$

$$5x = 31 - 17$$
$$5x = 14$$

$$x = \frac{14}{5} = 2.8$$

7. **(C)**—Watch what happens when you express everything as powers of 3:

$$3^{x^2}(9^x)3 = 27$$

$$3^{x^2}(3^{2x})3^1 = 3^3$$

The left side of the equation is the product of powers with the same base, so just add the exponents:

$$3^{x^2}(3^{2x})3^1 = 3^3$$

$$3^{x^2 + 2x + 1} = 3^3$$

Now the two sides of the equation are powers with the same base, so you can just set the exponents equal:

$$3^{x^2 + 2x + 1} = 3^3$$

$$x^2 + 2x + 1 = 3$$

$$(x + 1)^2 = 3$$

$$x + 1 = \pm\sqrt{3}$$

The positive value is $-1 + \sqrt{3}$, which is approximately 0.732.

8. **(D)**—First multiply both sides by $y - 4a$ to clear the denominator:

$$x = \frac{y + a^2}{y - 4a}$$

$$x(y - 4a) = y + a^2$$

$$xy - 4ax = y + a^2$$

Now move all terms with y to the left and all terms without y to the right:

$$xy - 4ax = y + a^2$$

$$xy - y = a^2 + 4ax$$

Now factor the left side and divide to isolate y :

$$xy - y = a^2 + 4ax$$

$$y(x - 1) = a^2 + 4ax$$

$$y = \frac{a^2 + 4ax}{x - 1}$$

9. **(B)**—With only two equations you won't be able to get numerical solutions for three unknowns. But apparently you can get far enough to rule out four of the five answer choices. How? Look for a way to combine the equations that leads somewhere useful. Notice that the first equation contains $+ky$ and the second equation contains $-ky$, so if you add the equations as they are, you'll lose those terms:

$$4x + ky = 15$$
$$\underline{x - ky = -25}$$
$$5x = -10$$
$$x = -2$$

There's not enough information to get numerical solutions for k or y, but you do know that $x = -2$, so the correct answer is the only choice that has an x-coordinate of -2.

10. **(D)**—If the absolute value of something is less than 8, then that something is between -8 and 8:

$$|4x + 3| < 8$$

$$-8 < 4x + 3 < 8$$

$$-11 < 4x < 5$$

$$-\frac{11}{4} < x < \frac{5}{4}$$

$$-2\frac{3}{4} < x < 1\frac{1}{4}$$

There are four integers in that range: $-2, -1, 0,$ and 1.

What's Next

Algebra is essential to a good score on any level SAT II: Mathematics Test. In the next chapter, we'll look at plane geometry, another topic that's crucial to your success, whether you're taking the Level IC or the Level IIC test.

CHAPTER 5

Plane Geometry

No matter which Mathematics Subject Test you're taking, you need what's in this chapter. According to the official breakdown, 20 percent of Level IC questions are plane geometry questions. But that's counting only the ones that are explicitly and primarily plane geometry questions. In fact, plane geometry is fundamental to solid geometry, coordinate geometry, and trigonometry. *The material in this chapter is relevant to nearly half the SAT II: Mathematics Level IC Test.*

Officially there are no plane geometry questions on Level IIC. But because plane geometry is fundamental to solid geometry, coordinate geometry, and trigonometry, *the material in this chapter is relevant to about 40 percent of the SAT II: Mathematics Level IIC Test.*

How to Use This Chapter

Maybe you already know all the plane geometry you need. You can find out by taking the Plane Geometry Diagnostic Test on page 67. The six plane geometry questions on the Diagnostic Test are typical of the Mathematics Subject Tests. Check your answers using the answer key following the test. No matter how you score, don't worry! The answer key also shows where to find a detailed explanation for each question. The "Find Your Study Plan" section that follows the test will suggest the next steps based on your performance on the Diagnostic.

 Find Your Level

How you use this chapter really depends on which test you're taking and how much time you have to prep. Find your level and pace below.

Taking the Mathematics Level IC Test? No matter how well you do on the Plane Geometry Diagnostic Test, read the rest of this chapter and do all the practice problems.

Plane Geometry Facts and Formulas in This Chapter

- Five Facts about Triangles (p. 72)
- Similar Figures (p. 73)
- Three Special Triangle Types (p. 74)
- Pythagorean Theorem (p. 75)
- Four Special Right Triangles (p. 75)
- Five Special Quadrilateral Types (p. 81)
- Polygon Angles (p. 83)
- Four Circle Formulas (p. 85)

Kaplan Strategies in This Chapter

- Mark up the figure (p. 70).
- Eyeball the figure (p. 74).
- Add to the figure (p. 78).
- Sketch a figure if none is provided (p. 82).

Mathematics Level IC Shortcut Take the Plane Geometry Diagnostic Test and check your answers. If you can answer at least four of the six questions correctly, then you already know the material in this chapter well enough to move on.

Taking the Mathematics Level IIC Test? No matter how well you do on the Plane Geometry Diagnostic Test, read the rest of this chapter and do all the practice problems.

Mathematics Level IIC Shortcut Take the Plane Geometry Diagnostic Test and check your answers. If you can answer at least four of the six questions correctly, then you already know the material in this chapter well enough to move on.

Panic Plan? Take the Plane Geometry Diagnostic Test and check your answers. The material in this chapter is vital. Don't try to move on to solid geometry, coordinate geometry, or trigonometry until you feel comfortable with the material in this chapter.

Plane Geometry Diagnostic Test

6 Questions (8 Minutes)

Directions: Solve the following problems and choose the best answer from those given. Fill in the oval corresponding to the best answer choice in the grid to the right of each question. (Answers are on page 70.)

DO YOUR FIGURING HERE.

1. In Figure 1, the length of segment PS is $2x + 12$, and the length of segment PQ is $6x - 10$. If R is the midpoint of segment QS, what is the length of segment PR ?

   (A) $-4x + 22$

   (B) $-2x + 11$

   (C) $-2x + 22$

   (D) $2x + 22$

   (E) $4x + 1$ Ⓐ Ⓑ Ⓒ Ⓓ Ⓔ

Figure 1

2. In Figure 2, $AB = BC$. If the area of $\triangle ABE$ is x, what is the area of $\triangle ACD$?

   (A) $x\sqrt{2}$

   (B) $x\sqrt{3}$

   (C) $2x$

   (D) $3x$

   (E) $4x$ Ⓐ Ⓑ Ⓒ Ⓓ Ⓔ

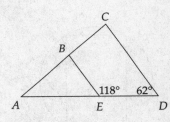

Figure 2

3. In Figure 3, $\triangle ABC$ is equilateral and $\triangle ADC$ is isosceles. If $AC = 1$, what is the distance from B to D ?

   (A) 0.286

   (B) 0.318

   (C) 0.333

   (D) 0.366

   (E) 0.383   Ⓐ Ⓑ Ⓒ Ⓓ Ⓔ

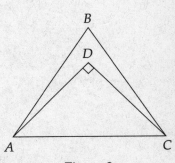

Figure 3

DO YOUR FIGURING HERE.

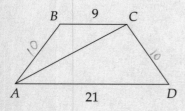

Figure 4

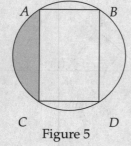

Figure 5

4. In Figure 4, the perimeter of isosceles trapezoid *ABCD* is 50. If *BC* = 9 and *AD* = 21, what is the length of diagonal *AC* ?

(A) 13

(B) 14

(C) 15

(D) 16

(E) 17

Ⓐ Ⓑ Ⓒ Ⓓ Ⓔ

5. A square and a regular hexagon have the same perimeter. If the area of the square is 2.25, what is the area of the hexagon?

(A) 2.250

(B) 2.598

(C) 2.838

(D) 3.464

(E) 3.375

Ⓐ Ⓑ Ⓒ Ⓓ Ⓔ

6. In Figure 5, rectangle *ABCD* is inscribed in a circle. If the radius of the circle is 1 and *AB* = 1, what is the area of the shaded region?

(A) 0.091

(B) 0.285

(C) 0.614

(D) 0.705

(E) 0.732

Ⓐ Ⓑ Ⓒ Ⓓ Ⓔ

STOP! END OF TEST. DO NOT TURN THE PAGE UNTIL YOU ARE READY TO CHECK YOUR ANSWERS.

Find Your Study Plan

The answer key on the following page shows where in this chapter to find explanations for the questions you missed. Here's how you should proceed based on your Diagnostic Test score.

6: Superb! You really know your plane geometry. Unless you have lots of time and just love to read about plane geometry, you might consider skipping this chapter. You seem to know it all already. Just to make absolutely sure, you could look over the plane geometry facts, formulas, and strategies in the margins of this chapter. And if all you want is more plane geometry questions to try, go to the Follow-Up Test at the end of this chapter.

4–5: Excellent! You're quite good at plane geometry. Some of these are especially difficult questions. If you're taking a "shortcut" or you're on the Panic Plan, you don't really have time to study this chapter, and you don't really need to. You might at least look over those pages that address the questions you didn't get right. You could also look over the plane geometry facts, formulas, and strategies in the margins of this chapter. If you just want to try more plane geometry questions, go to the Follow-Up Test at the end of the chapter.

0–3: You should read this chapter and do the Follow-Up Test at the end. You need to have a good command of the material in this chapter before moving on to later chapters on solid geometry, coordinate geometry, and trigonometry.

Test Topics

This icon appears next to each discussion of a math topic that's tested on the SAT II.

Plane Geometry Test Topics

The questions in the Plane Geometry Diagnostic Test are typical of the SAT II: Mathematics Level IC Test. They range from segments, through polygons (especially triangles), to circles and multiple figures. None of these questions would appear on the Mathematics Level IIC Test, but that's only because the Level IIC test makers assume you know plane geometry cold. To perform well on Level IIC solid geometry, coordinate geometry, and trigonometry questions, you must first be able to do the kind of plane geometry questions you'll see here. In this chapter we'll use these questions to review the plane geometry you're expected to know for the Mathematics Subject Tests. We will also use these questions to demonstrate some effective problem-solving techniques, alternative methods, and test-taking strategies that apply to SAT II plane geometry questions.

Mark Up the Figure.

Try to put all the given information into the figure so that you can see everything at a glance and don't have to keep going back and forth between the question stem and the figure.

 Adding and Subtracting Segment Lengths

The simplest type of plane geometry questions you may encounter on the SAT II are like Example 1, which involves adding and subtracting segment lengths. It's typical of the SAT II that the lengths you have to add and subtract are algebraic expressions rather than numbers:

Example 1

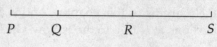

Figure 1

In Figure 1, the length of segment PS is $2x + 12$, and the length of segment PQ is $6x - 10$. If R is the midpoint of segment QS, what is the length of segment PR ?

(A) $-4x + 22$

(B) $-2x + 11$

(C) $-2x + 22$

(D) $2x + 22$

(E) $4x + 1$

Usually the best thing to do to start on a plane geometry question is to *mark up the figure*. Put as much of the information into the figure as you can. That's a good way to organize your thoughts. And that way you don't have to go back and forth between the figure and the question.

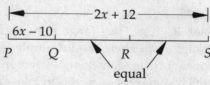

Now you can plan your attack. First subtract PQ from the whole length PS to get QS :

$$QS = PS - PQ = (2x + 12) - (6x - 10) = 2x + 12 - 6x + 10 = -4x + 22$$

Then, because R is the midpoint of QS, you can divide QS by 2 to get QR and RS :

$$QR = RS = \frac{QS}{2} = \frac{-4x + 22}{2} = -2x + 11$$

Watch out! That matches choice (B), but it's the answer to the wrong question. What you're looking for is PR, so you have to add:

$$PR = PQ + QR = (6x - 10) + (-2x + 11) = 4x + 1$$

The answer is (E).

 Basic Traits of Triangles

Most SAT II plane geometry questions are about closed figures: polygons and circles. And the test makers' favorite closed figure by far is the three-sided polygon, that is, the triangle. All three-sided polygons are interesting because they share so many characteristics, and certain *special* three-sided polygons—equilateral, isosceles, and right triangles—are interesting because of their special characteristics.

Let's look at the traits that all triangles share.

Sum of the interior angles: The three interior angles of any triangle add up to 180°.

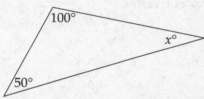

In the figure above, $x + 50 + 100 = 180$, so $x = 30$.

Measure of an exterior angle: The measure of an exterior angle of a triangle is equal to the sum of the measures of the remote interior angles.

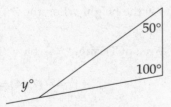

In the figure above, the measure of the exterior angle labeled $y°$ is equal to the sum of the measures of the remote interior angles: $y = 50 + 100 = 150$.

Five Facts about Triangles

1. Interior angles add up to 180°.

2. Exterior angle equals sum of remote interior angles.

3. Exterior angles add up to 360°.

4. Area of a triangle = $\frac{1}{2}$(base)(height).

5. Each side is greater than the difference and less than the sum of the other two sides.

Sum of the exterior angles: The measures of the three exterior angles of any triangle add up to 360°.

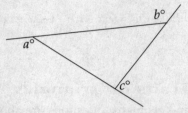

In the figure above, $a + b + c = 360$. (*Note:* In fact, the measures of the exterior angles of *any polygon* add up to 360°.)

Area formula: The general formula for the area of a triangle is always the same. The formula is:

$$\text{Area of Triangle} = \frac{1}{2}(\text{base})(\text{height})$$

The height is the perpendicular distance between the side that's chosen as the base and the opposite vertex.

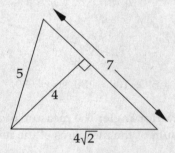

In the triangle above, 4 is the height when the 7 side is chosen as the base.

$$\text{Area of Triangle} = \frac{1}{2}(\text{base})(\text{height})$$
$$= \frac{1}{2}(7)(4) = 14$$

Triangle Inequality Theorem: The length of any one side of a triangle must be greater than the positive difference and less than the sum of the lengths of the other two sides. For example, if it is given that the length of one side is 3 and the length of another side is 7, then the length of the third side must be greater than $7 - 3 = 4$ and less than $7 + 3 = 10$.

Similar Triangles

In Example 2 you're asked to express the area of one triangle in terms of the area of another.

Example 2

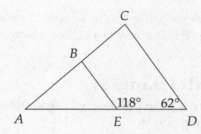

Figure 2

In Figure 2, $AB = BC$. If the area of $\triangle ABE$ is x, what is the area of $\triangle ACD$?

(A) $x\sqrt{2}$

(B) $x\sqrt{3}$

(C) $2x$

(D) $3x$

(E) $4x$

You might wonder here how you're supposed to find the area of $\triangle ACD$ when you're given no lengths you can use for a base or an altitude. The only numbers you have are the angle measures. They must be there for some reason—the test makers rarely provide superfluous information. In fact, because the two angle measures provided add up to 180°, they tell you that BE and CD are parallel. And that, in turn, tells you that $\triangle ABE$ is *similar* to $\triangle ACD$—because they have the same three angles.

Similar triangles are triangles that have the same shape: Corresponding angles are equal and corresponding sides are proportional. In this case, because it's given that $AB = BC$, you know that AC is twice AB and that corresponding sides are in a ratio of 2:1. Each side of the larger triangle is twice the length of the corresponding side of the smaller triangle. That doesn't mean, however, that the ratio of the *areas* is also 2:1. In fact, the area ratio is the *square* of the side ratio, and the larger triangle has *four times* the area of the smaller triangle, so the answer is (E).

Alternative method If you didn't see the similar triangles, or if you didn't know for sure how the area of the larger triangle is related to the area

 Similar Figures

If you double the lengths of all the sides of a polygon, you more than double the area of the polygon. In fact, you quadruple it. The area ratio between similar figures is the square of the side ratio.

Eyeball the Figure.

You can assume that a figure is drawn to scale unless the problem says otherwise. So, when you're stuck on a geometry question and don't know what else to do, see if you can at least use your eyes to eliminate a few answer choices as visibly too small or too big.

Three Special Triangle Types

1. **Isosceles:** two equal sides and two equal angles.
2. **Equilateral:** three equal sides and three 60° angles. If the length of one side is *s*, then:

   Area of Equilateral Triangle
   $$=\frac{s^2\sqrt{3}}{4}$$

3. **Right:** one right angle. You can use the legs to find the area:
   Area of Right Triangle =
   $$\frac{1}{2}(\text{leg}_1)(\text{leg}_2)$$

of the smaller triangle, you could have at least eliminated some answer choices based on appearances. Look at the figure and *use your eyes* to compare the areas. (We call this method *eyeballing*.) Doesn't it look like the larger triangle has more than twice as much room inside it as the smaller triangle? That means that answer choices (A), (B), and (C) are all visibly too small. If you can narrow the choices down to two, it certainly pays to guess.

Eyeballing is never what you're *supposed* to do to answer a question, but if you don't see a better way, eyeballing's better than skipping.

Special Triangles

Three special triangle types deserve extra attention:
- Isosceles triangles
- Equilateral triangles
- Right triangles

Be sure you know not just the definitions of these triangle types, but more importantly their special characteristics: side relationships, angle relationships, and area formulas.

Isosceles triangle: An isosceles triangle is a triangle that has *two equal sides.* Not only are two sides equal, but the angles opposite the equal sides, called base angles, are also equal.

Equilateral triangle: An equilateral triangle is a triangle that has *three equal sides.* Since all the sides are equal, all the angles are also equal. All three angles in an equilateral triangle measure 60 degrees, regardless of the lengths of the sides. You can find the area of an equilateral triangle by dividing it into two 30-60-90 triangles, or you can use this formula in terms of the length of one side *s* :

$$\text{Area of Equilateral Triangle} = \frac{s^2\sqrt{3}}{4}$$

Right triangle: A right triangle is a triangle with a right angle. The two sides that form the right angle are called *legs*, and you can use them as the base and height to find the area of a right triangle.

$$\text{Area of Right Triangle} = \frac{1}{2}(\text{leg}_1)(\text{leg}_2)$$

Pythagorean theorem: If you know any two sides of a right triangle, you can find the third side by using the Pythagorean theorem:

$$(leg_1)^2 + (leg_2)^2 = (hypotenuse)^2$$

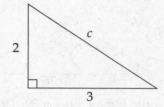

For example, if one leg is 2 and the other leg is 3, then:

$$2^2 + 3^2 = c^2$$
$$c^2 = 4 + 9$$
$$c = \sqrt{13}$$

Pythagorean triplet: A Pythagorean triplet is a set of integers that fit the Pythagorean theorem. The simplest Pythagorean triplet is (3, 4, 5). In fact, any integers in a 3:4:5 ratio make up a Pythagorean triplet. And there are many other Pythagorean triplets: (5, 12, 13); (7, 24, 25); (8, 15, 17); (9, 40, 41); all their multiples; and infinitely many more.

3-4-5 triangle: If a right triangle's leg-to-leg ratio is 3:4, or if the leg-to-hypotenuse ratio is 3:5 or 4:5, then it's a 3-4-5 triangle and you don't need to use the Pythagorean theorem to find the third side. Just figure out what multiple of 3-4-5 it is.

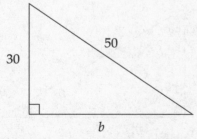

In the right triangle shown, one leg is 30 and the hypotenuse is 50. This is 10 times 3-4-5. The other leg is 40.

Pythagorean Theorem

For all right triangles:
$$(leg_1)^2 + (leg_2)^2 = (hypotenuse)^2$$

Four Special Right Triangles

1. The 3-4-5 Triangle

Four Special Right Triangles (continued)

2. The 5-12-13 Triangle

3. The 45-45-90 Triangle

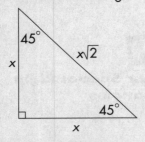

4. The 30-60-90 Triangle

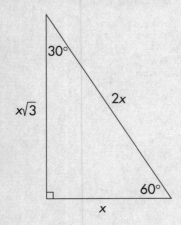

5-12-13 triangles: If a right triangle's leg-to-leg ratio is 5:12, or if the leg-to-hypotenuse ratio is 5:13 or 12:13, then it's a 5-12-13 triangle and you don't need to use the Pythagorean theorem to find the third side. Just figure out what multiple of 5-12-13 it is.

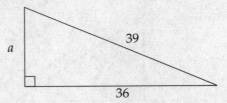

Here one leg is 36 and the hypotenuse is 39. This is 3 times 5-12-13. The other leg is 3×5 or 15.

45-45-90 triangles: The sides of a 45-45-90 triangle are in a ratio of $1:1:\sqrt{2}$.

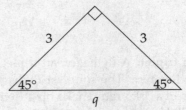

If one leg is 3, then the other leg is also 3, and the hypotenuse is equal to a leg times $\sqrt{2}$, or $3\sqrt{2}$.

30-60-90 triangles: The sides of a 30-60-90 triangle are in a ratio of $1:\sqrt{3}:2$. You don't need to use the Pythagorean theorem.

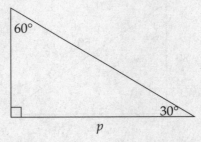

If the hypotenuse is 6, then the shorter leg is half that, or 3; and then the longer leg is equal to the short leg times $\sqrt{3}$, or $3\sqrt{3}$.

Example 3 includes one triangle that's equilateral and another that's both right and isosceles.

Example 3

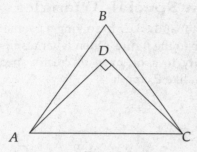

Figure 3

In Figure 3, $\triangle ABC$ is equilateral and $\triangle ADC$ is isosceles. If $AC = 1$, what is the distance from B to D ?

(A) 0.286

(B) 0.318

(C) 0.333

(D) 0.366

(E) 0.383

To get the answer to this question, you need to know about equilateral, 45-45-90, and 30-60-90 triangles. If you drop an altitude from B through D, you will divide the equilateral triangle into two 30-60-90 triangles and you will divide the right isosceles (or 45-45-90) triangle into two smaller right isosceles triangles:

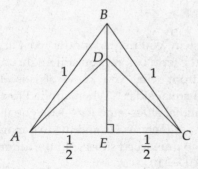

Using the side ratios for 30-60-90s and 45-45-90s, you know that

$BE = \dfrac{\sqrt{3}}{2}$ and that $DE = \dfrac{1}{2}$.

Therefore, $BD = BE - DE = \dfrac{\sqrt{3}}{2} - \dfrac{1}{2} = \dfrac{\sqrt{3}-1}{2} \approx \dfrac{1.732-1}{2} = 0.366$.

The answer is (D).

Add to the Figure.

Often the breakthrough on a plane geometry problem comes when you add a line segment or two to the figure. Perpendiculars can be especially useful. They can function as rectangle sides or triangle altitudes or right triangle legs.

 Hidden Special Triangles

It happens a lot that the key to solving a geometry problem is to add a line segment or two to the figure. Often what results is one or more special triangles. The ability to spot, even to create, special triangles comes in handy in a question like Example 4.

Example 4

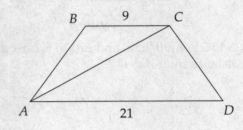

Figure 4

In Figure 4, the perimeter of isosceles trapezoid $ABCD$ is 50. If $BC = 9$ and $AD = 21$, what is the length of diagonal AC ?

(A) 13

(B) 14

(C) 15

(D) 16

(E) 17

As you read the stem, you might wonder what an *isosceles trapezoid* is. If you'd never heard the term before, you still might have been able to extrapolate its meaning from what you know of isosceles triangles. Isosceles means "having two equal sides." When applied to a trapezoid, it tells you that the two nonparallel sides—the legs—are equal. In this case that's AB and CD. If the total perimeter is 50, and the two marked sides add up to 21 + 9 = 30, then the two unmarked sides split the difference of 50 − 30 = 20. In other words, $AB = CD = 10$.

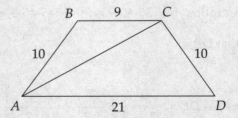

There aren't any special triangles yet. As so often happens, though, you can get some by constructing altitudes. Drop perpendiculars from points *B* and *C* and you make two right triangles. The length 21 of side *AD* then gets split into 6, 9, and 6.

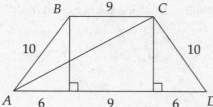

Now you can see that those right triangles are 3-4-5s (times 2) and that the height of the trapezoid is 8. Now look at the right triangle of which *AC* is the hypotenuse.

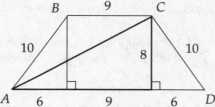

One leg is 6 + 9 = 15, and the other leg is 8; therefore the hypotenuse *AC* is as follows.

$$\text{hypotenuse} = \sqrt{(\text{leg}_1)^2 + (\text{leg}_2)^2}$$
$$= \sqrt{15^2 + 8^2}$$
$$= \sqrt{225 + 64} = \sqrt{289} = 17$$

So *AC* = 17 and the answer is (E).

Special Quadrilaterals

The trapezoid is just one of five special quadrilaterals you need to be familiar with. As with triangles, there is some overlap among these categories, and there are some figures that fit into none of these categories. Just as a 45-45-90 triangle is both right and isosceles, a quadrilateral with four equal sides and four right angles is not only a square, but also a rhombus, a rectangle, and a parallelogram. It is wise to have a solid grasp of the definitions and special characteristics of these five quadrilateral types.

Trapezoids: A trapezoid is a four-sided figure with one pair of parallel sides and one pair of nonparallel sides.

$$\text{Area of Trapezoid} = \left(\frac{\text{base}_1 + \text{base}_2}{2}\right) \times \text{height}$$

Think of this formula as the average of the bases (the two parallel sides) times the height (the length of the perpendicular altitude).

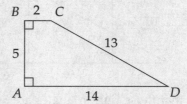

In the trapezoid *ABCD* above, you can use side *AD* for the height. The average of the bases is $\frac{2+14}{2} = 8$, so the area is 8×5, or 40.

Parallelograms: A parallelogram is a four-sided figure with two pairs of parallel sides. Opposite sides are equal. Opposite angles are equal. Consecutive angles add up to 180°.

Area of Parallelogram = base × height

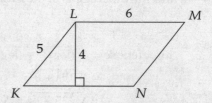

In parallelogram *KLMN* above, 4 is the height when *LM* or *KN* is used as the base. Base × height = $6 \times 4 = 24$.

Remember that to find the area of a parallelogram you need the height, which is the perpendicular distance from the base to the opposite side. You can use a side of a parallelogram for the height only when the side is perpendicular to the base, in which case you have a rectangle.

Rectangles: A rectangle is a four-sided figure with four right angles. Opposite sides are equal. Diagonals are equal. The perimeter of a rectangle is equal to the sum of the lengths of the four sides, which is equal to 2(length + width).

Area of Rectangle = length × width

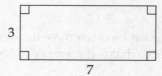

The area of a 7-by-3 rectangle is $7 \times 3 = 21$.

Rhombus: A rhombus is a four-sided figure with four equal sides.

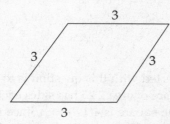

All four sides of the quadrilateral above have the same length, so it's a rhombus. A rhombus is also a parallelogram, so to find the area of a rhombus, you need its height. The more a rhombus "leans over," the smaller the height and therefore the smaller the area. The maximum area for a rhombus of a certain perimeter is that rhombus that has each pair of adjacent sides perpendicular, in which case you have a square.

Square: A square is a four-sided figure with four right angles and four equal sides. A square is also a rectangle, a parallelogram, and a rhombus. The perimeter of a square is equal to 4 times the length of one side.

$$\text{Area of Square} = (\text{side})_2$$

The square above, with sides of length 5, has an area of $5^2 = 25$.

 ## Polygons—Perimeter and Area

The SAT II: Mathematics test makers like to write problems that combine the concepts of perimeter and area. What you need to remember is that perimeter and area are not directly related. In Example 5, for instance, you have two figures with the same perimeter, but that doesn't mean they have the same area.

Five Special Quadrilateral Types

1. **Trapezoid:** one pair of parallel sides.

    Area of Trapezoid =
    $$\left(\frac{\text{base}_1 + \text{base}_2}{2}\right) \times \text{height}$$

2. **Parallelogram:** two pairs of parallel sides.

    Area of Parallelogram = base × height

3. **Rectangle:** four right angles.

    Perimeter of Rectangle = 2(length + width)

    Area of Rectangle = length × width

4. **Rhombus:** four equal sides.

    Perimeter of Rhombus = 4 × side

    Area of Rhombus = base × height

5. **Square:** four right angles and four equal sides.

    Perimeter of Square = 4 × side

    Area of Square = $(\text{side})^2$

Sketch a Figure If None Is Provided.

The best way to get a handle on a figureless geometry problem is usually to sketch a figure of your own. You don't have to be an artist. Just be neat and clear enough to get the picture.

Example 5

A square and a regular hexagon have the same perimeter. If the area of the square is 2.25, what is the area of the hexagon?

(A) 2.250

(B) 2.598

(C) 2.838

(D) 3.464

(E) 3.375

The way to get started with this question is to sketch what's described in the question. A square of area 2.25 has sides each of length $\sqrt{2.25} = 1.5$. So the perimeter of the square is 4(1.5) = 6. Since that's also the perimeter of the regular hexagon, and a regular hexagon has six equal sides, the length of each side of the hexagon is 1.

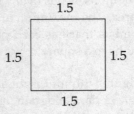
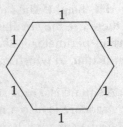

Now the problem is one of finding the area of a regular hexagon of side length 1. The fastest way to do that would be to use the formula, if you know it. If the length of one side is s :

$$\textbf{Area of Hexagon} = \frac{3s^2\sqrt{3}}{2}$$

This formula is not one the test makers expect you to know—there's always a way around it—but if you like formulas and you're good at memorizing them, it can only help. Let's proceed, however, as if we didn't know the formula. Another way to go about finding this area is to add a line segment or two to the figure and divide it up into more familiar shapes. You could, for example, draw in three diagonals and turn the hexagon into six equilateral triangles of side 1:

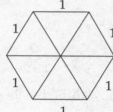

Each of those six triangles has base 1 and height $\dfrac{\sqrt{3}}{2}$, and therefore:

$$\text{Area of one triangle} = \frac{1}{2}(\text{base})(\text{height}) = \frac{1}{2}(1)\left(\frac{\sqrt{3}}{2}\right) = \frac{\sqrt{3}}{4}$$

The area of the hexagon is 6 times that.

$$\text{Area of hexagon} = 6\left(\frac{\sqrt{3}}{4}\right) = \frac{3\sqrt{3}}{2} \approx 2.598$$

The answer is (B).

 ## Circles—Four Formulas

After the triangle, the test makers' favorite plane geometry figure is the circle. Circles don't come in as many varieties as triangles do. In fact, all circles are similar—they're all the same shape. The only difference among them is size. So you don't have to learn to recognize types or remember names. All you have to know about circles is how to find four things:

- Circumference
- Length of an arc
- Area
- Area of a sector

You could think of the task as one of memorizing four formulas, but you'll be better off in the end if you have some idea of where the arc and sector formulas come from and how they are related to the circumference and area formulas.

Circumference: Circumference is a measurement of length. You could think of it as the perimeter: It's the total distance around the circle. If the radius of the circle is r:

Circumference = $2\pi r$

Polygon Angles

For any polygon of n sides:

Sum of Interior Angles = $(n-2) \times 180°$

Sum of Exterior Angles = $360°$

Since the diameter is twice the radius, you can easily express the formula in terms of the diameter d :

Circumference = πd

In the circle above, the radius is 3, and so the circumference is $2\pi(3) = 6\pi$.

Length of an arc: An arc is a piece of the circumference. If n is the degree measure of the arc's central angle, then the formula is:

Length of an Arc = $\left(\dfrac{n}{360}\right)(2\pi r)$

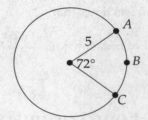

In the figure above, the radius is 5 and the measure of the central angle is 72°. The arc length is $\dfrac{72}{360}$ or $\dfrac{1}{5}$ of the circumference:

$$\left(\frac{72}{360}\right)(2\pi)(5) = \left(\frac{1}{5}\right)(10\pi) = 2\pi$$

Area: The area of a circle is usually found using this formula in terms of the radius r :

$$\text{Area of a Circle} = \pi r^2$$

The area of the circle above is $\pi(4)^2 = 16\pi$.

Area of a sector: A sector is a piece of the area of a circle. If n is the degree measure of the sector's central angle, then the area formula is:

$$\text{Area of a Sector} = \left(\frac{n}{360}\right)\left(\pi r^2\right)$$

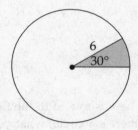

In the figure above, the radius is 6 and the measure of the sector's central angle is 30°. The sector has $\dfrac{30}{360}$ or $\dfrac{1}{12}$ of the area of the circle:

$$\left(\frac{30}{360}\right)(\pi)\left(6^2\right) = \left(\frac{1}{12}\right)(36\pi) = 3\pi$$

Four Circle Formulas

Circumference $= 2\pi r = \pi d$

Length of an Arc =

$$\left(\frac{n}{360}\right)\left(2\pi r\right)$$

Area of a Circle $= \pi r^2$

Area of a Sector =

$$\left(\frac{n}{360}\right)\left(\pi r^2\right)$$

Circles Combined with Other Figures

Some of the most challenging plane geometry questions are those that combine circles with other figures, like Example 6:

Example 6

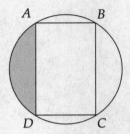

Figure 5

In Figure 5, rectangle *ABCD* is inscribed in a circle. If the radius of the circle is 1 and *AB* = 1, what is the area of the shaded region?

(A) 0.091

(B) 0.285

(C) 0.614

(D) 0.705

(E) 0.732

Once again, the key here is to add to the figure. And in this case, as is so often the case when there's a circle, what you should add is radii. The equilateral triangles tell you that the central angles are 60° and 120°.

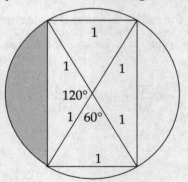

The shaded region is what's left of the 120° sector after you subtract the triangle with the 120° vertex angle. To find the area of the shaded region, you want to find the areas of the sector and triangle, and then subtract. The sector is exactly one-third of the circle (because 120° is one-third of 360°), and so:

$$\text{Area of sector} = \frac{1}{3}\pi r^2 = \frac{1}{3}\pi(1)^2 = \frac{\pi}{3} \approx 1.047$$

You can divide the triangle into two 30-60-90s:

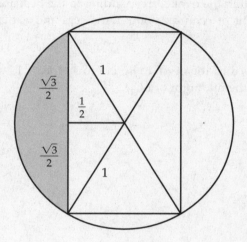

The area of each 30-60-90 triangle is $\frac{1}{2}\left(\frac{1}{2}\right)\left(\frac{\sqrt{3}}{2}\right) = \frac{\sqrt{3}}{8}$, so the area of the

triangle with the 120° vertex is twice that, or $\frac{\sqrt{3}}{4} \approx 0.433$. The shaded area,

then, is about $1.047 - 0.433 = 0.614$. The answer is (C).

Now that you've reviewed all the relevant plane geometry facts and formulas, seen some of the test makers' favorite plane geometry situations, and learned a few good Kaplan strategies, it's time to put yourself to the test again. If you were disappointed in your performance on the Plane Geometry Diagnostic Test, here's your chance to make up for it.

Plane Geometry Follow-Up Test

6 Questions (8 Minutes)

Directions: Solve the following problems and choose the best answer from those given. Fill in the oval corresponding to the best answer choice in the grid to the right of each question. (Answers and explanations begin on page 93.)

DO YOUR FIGURING HERE.

1. In Figure 1, the ratio of AB to BC is 7 to 5. If $AC = 1$, what is the distance from A to the midpoint of BC ?

    (A) $\dfrac{5}{8}$

    (B) $\dfrac{2}{3}$

    (C) $\dfrac{17}{24}$

    (D) $\dfrac{3}{4}$

    (E) $\dfrac{19}{24}$

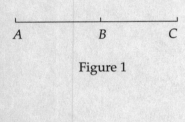

Figure 1

2. In $\triangle PRS$ in Figure 2, RT is the altitude to side PS, and QS is the altitude to side PR. If $RT = 7$, $PR = 8$, and $QS = 9$, what is the length of side PS ?

    (A)   5.14

    (B)   6.22

    (C)   7.87

    (D)   10.29

    (E)   13.44

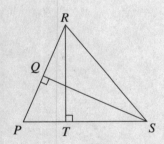

Figure 2

DO YOUR FIGURING HERE.

3. In Figure 3, *QS* and *PT* are parallel, and the lengths of segments *PQ* and *QR* are as marked. If the area of $\triangle QRS$ is *x*, what is the area of $\triangle PRT$ in terms of *x* ?

(A) $\dfrac{3x}{2}$

(B) $\dfrac{9x}{4}$

(C) $\dfrac{5x}{2}$

(D) $3x$

(E) $\dfrac{25x}{4}$

 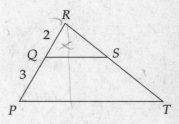

Ⓐ Ⓑ Ⓒ Ⓓ Ⓔ

Figure 3

4. In Figure 4, if *AB* = 2, what is the area of $\triangle ABC$?

(A) 2.45

(B) 2.73

(C) 3.86

(D) 4.89

(E) 5.46

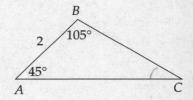

Ⓐ Ⓑ Ⓒ Ⓓ Ⓔ

Figure 4

5. In Figure 5, the area of parallelogram *PQRS* is 24 and *QR* = 6. If *QT* is perpendicular to *PS* and if *T* is the midpoint of *PS*, what is the perimeter of *PQRS* ?

(A) 20

(B) 22

(C) 24

(D) 26

(E) 28

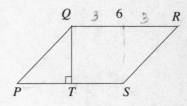

Ⓐ Ⓑ Ⓒ Ⓓ Ⓔ

Figure 5

DO YOUR FIGURING HERE.

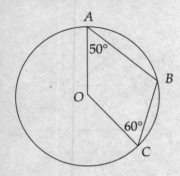

Figure 6

6. In Figure 6, points *A*, *B*, and *C* lie on the circumference of the circle centered at *O*. If ∠*OAB* measures 50° and ∠*BCO* measures 60°, what is the measure of ∠*AOC* ?

(A) 110°

(B) 120°

(C) 130°

(D) 140°

(E) 150°

Ⓐ Ⓑ Ⓒ Ⓓ Ⓔ

STOP!   **END OF TEST. DO NOT TURN THE PAGE UNTIL YOU ARE READY TO CHECK YOUR ANSWERS.**

Turn the page
for answers and explanations
to the Follow-Up Test.

Follow-Up Test—Answers and Explanations

Answer Key   1. *E*   2. *D*   3. *E*   4. *B*   5. *B*   6. *D*

1. **(E)**—Mark up the figure:

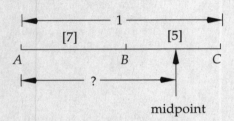

midpoint

The brackets around the 7 and 5 are meant to show that they're proportions, not actual lengths. Since the ratio of *AB* to *BC* is 7 to 5, you can say that 7 parts out of 12 are in *AB* and 5 parts out of 12 are in *BC*. And because the whole length *AC* is 1, $AB = \frac{7}{12}$ and $BC = \frac{5}{12}$. The midpoint of *BC* divides it in half, so each half-length is half of $\frac{5}{12}$, which is $\frac{5}{24}$. The length you're looking for is $\frac{7}{12} + \frac{5}{24} = \frac{14}{24} + \frac{5}{24} = \frac{19}{24}$.

2. **(D)**—Put the given lengths into the figure:

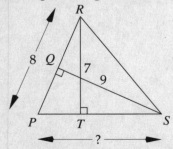

The key is realizing that whatever base-altitude pair you take for the same triangle, you'll get the same area. One-half the product of *PS* and *RT* is the same as one-half the product of *PR* and *QS*. Or, ignoring the one-halfs, you can just say that the products are equal.

$$\frac{1}{2}(PS)(RT) = \frac{1}{2}(PR)(QS)$$
$$(PS)(RT) = (PR)(QS)$$
$$(PS)(7) = (8)(9)$$
$$PS = \frac{(8)(9)}{7} = \frac{72}{7} \approx 10.29$$

3. **(E)**—The only information that's in the question and not in the figure is that *QS* and *PT* are parallel and that the area of △*QRS* is *x*. That *QS* and *PT* are parallel tells you that triangles *PRT* and *QRS* are similar—they have the same angles. Because the triangles are similar, the sides are proportional. Because the ratio of *PR* to *QR* is $\frac{5}{2}$, the ratio of any pair of corresponding sides will also be $\frac{5}{2}$. But that's not the ratio of the areas. Remember that the area ratio between similar figures is the *square* of the side ratio. Here the side ratio is $\frac{5}{2}$, so the area ratio is $\left(\frac{5}{2}\right)^2 = \frac{25}{4}$. If the area of the small triangle is *x*, then the area of the large one is $\frac{25x}{4}$.

4. **(B)**—Drop an altitude and you'll reveal two hidden special triangles:

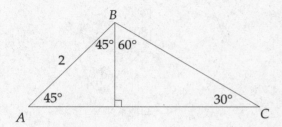

Now you can use the known side ratios of 45-45-90 and 30-60-90 triangles to get all the lengths you need. If the hypotenuse of a 45-45-90 is 2, then each leg is $\frac{2}{\sqrt{2}} = \sqrt{2}$:

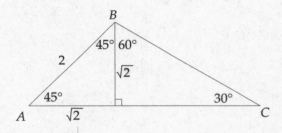

Now that you know the short leg of the 30-60-90, you can multiply that by $\sqrt{3}$ to find that the longer leg is $\sqrt{6}$:

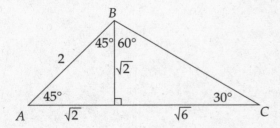

Now you have the base and the height. The base is $\sqrt{2} + \sqrt{6}$, and the height is $\sqrt{2}$, so:

Area of a triangle $= \frac{1}{2}$ (base)(hight)

$$= \frac{1}{2}(\sqrt{2} + \sqrt{6})(\sqrt{2}) = \frac{1}{2}(2 + \sqrt{12})$$

$$= \frac{1}{2}(2 + 2\sqrt{3}) = 1 + \sqrt{3} \approx 2.73$$

5. **(B)**—Because *PQRS* is a parallelogram, opposite sides are equal and *PS* = 6. Midpoint *T* divides that 6 into two 3s. The area of the parallelogram, which is equal to the base times the height, is given as 24, and because the base is 6, altitude *QT* must be 4. Put all this into the figure and the hidden special triangle reveals itself.

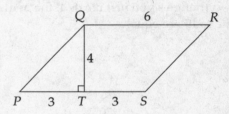

Now you can see that Δ*PQT* is a 3-4-5 right triangle and that *PQ* = 5. Opposite sides are equal, so *RS* = 5, too, and the perimeter is 5 + 6 + 5 + 6 = 22.

6. (D)—As is so often true when circles are combined with other figures, the key to solving this question is to draw in a radius:

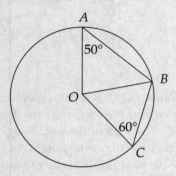

All the radii of a circle are equal, so within each of the two triangles you just created, the angles opposite the radii are equal:

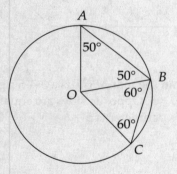

Now that you know two angles in each triangle, you can figure out the third:

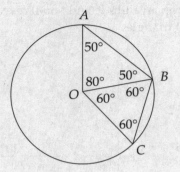

The angle you're looking for measures 80° + 60° = 140°.

What's Next

In the next chapter, we'll shift gears from plane geometry to solid geometry. But make certain that you've mastered the principles in this chapter before moving on! Plane geometry skills will help assure a good score on the SAT II: Mathematics Tests.

CHAPTER 6

Solid Geometry

The material in this chapter is relevant to both Mathematics Subject tests. Level IC and Level IIC both include a few solid geometry questions. But this is not a big and vital category on either test. A typical Level IC has three solid geometry questions; a typical Level IIC has four.

How to Use This Chapter

Maybe you already know all the solid geometry you need. You can find out by taking the Solid Geometry Diagnostic Test on page 98. The five solid geometry questions on the Diagnostic Test are typical of the Mathematics Subject Tests. Check your answers using the answer key following the test. No matter how you score, don't worry! The answer key also shows where to find a detailed explanation for each question. The "Find Your Study Plan" section that follows the test will suggest next steps based on your performance on the Diagnostic.

 Find Your Level

How you use this chapter really depends on which test you're taking and how much time you have to prep. Find your level and pace below.

Taking the Mathematics Level IC Test? Take the Solid Geometry Diagnostic Test to find out how much you already know about solid geometry. Then read the rest of the chapter and do the Follow-Up Test at the end.

Mathematics Level IC Shortcut Take the Solid Geometry Diagnostic Test and check your answers. If you can answer most of the questions correctly, you should skip this chapter.

Taking the Mathematics Level IIC Test? Take the Solid Geometry Diagnostic Test to find out how much you already know about solid geometry. Then read the rest of the chapter and do the Follow-Up Test at the end.

Solid Geometry Facts and Formulas in This Chapter

- Five Formulas You Don't Need to Memorize (p. 103)
- Surface Area of a Rectangular Solid (p. 105)
- Distance between Opposite Vertices of a Rectangular Solid (p. 107)
- Volume Formulas for Uniform Solids (p. 108)

Kaplan Strategies in This Chapter

- Pick numbers (p. 104).
- Brush up on your algebra (p. 105).

Mathematics Level IIC Shortcut Take the Solid Geometry Diagnostic Test and check your answers. If you can answer most of the questions correctly, you should skip this chapter.

Panic Plan? Skip this chapter.

Solid Geometry Diagnostic Test

---

5 Questions (6 Minutes)

Directions: Solve the following problems and choose the best answer from those given. Fill in the oval corresponding to the best answer choice in the grid to the right of each question. (Answers are on page 102.)

> **Reference Information:** Use the following formulas as needed.
>
> **Right circular cone:** If r = radius and h = height, then **Volume** $= \dfrac{1}{3}\pi r^2 h$; and if c = circumference of the base and ℓ = slant height, then **Lateral Area** $= \dfrac{1}{2}c\ell$.
>
> **Sphere:** If r = radius, then **Volume** $= \dfrac{4}{3}\pi r^3$ and **Surface Area** $= 4\pi r^2$.
>
> **Pyramid:** If B = area of the base and h = height, then **Volume** $= \dfrac{1}{3}Bh$.

DO YOUR FIGURING HERE.

1. A right circular cone and a sphere have equal volumes. If the radius of the base of the cone is $2x$ and the radius of the sphere is $3x$, what is the height of the cone in terms of x ?

    (A) x

    (B) $\dfrac{3x}{2}$

    (C) $\dfrac{4x}{3}$

    (D) $20x$

    (E) $27x$

    Ⓐ Ⓑ Ⓒ Ⓓ Ⓔ

2. In the rectangular solid in Figure 1, what is the distance from vertex A to vertex B ?

    (A) $\sqrt{65}$

    (B) $7\sqrt{2}$

    (C) 9

    (D) 11

    (E) 15

    Ⓐ Ⓑ Ⓒ Ⓓ Ⓔ

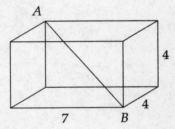

Figure 1

3. In Figure 2, the bases of the right uniform solid are triangles with sides of lengths 3, 4, and 5. If the volume of the solid is 30, what is the total surface area?

    (A) 17

    (B) 30

    (C) 36

    (D) 60

    (E) 72

    Ⓐ Ⓑ Ⓒ Ⓓ Ⓔ

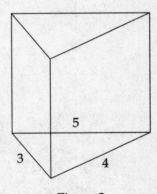

Figure 2

99

DO YOUR FIGURING HERE.

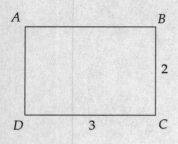

Figure 3

4. If the rectangle in Figure 3 is rotated 360° about side *BC*, what is the volume of the resulting solid?

    (A) 12.00

    (B) 18.00

    (C) 28.27

    (D) 37.70

    (E) 56.55

5. What is the maximum number of rectangular blocks, each with dimensions 3 centimeters by 5 centimeters by 7 centimeters, that can be placed inside a rectangular carton with inside dimensions 14 centimeters by 15 centimeters by 16 centimeters ?

    (A) 30

    (B) 32

    (C) 33

    (D) 35

    (E) 45

    Ⓐ Ⓑ Ⓒ Ⓓ Ⓔ

STOP! **END OF TEST. DO NOT TURN THE PAGE UNTIL YOU ARE READY TO CHECK YOUR ANSWERS.**

Find Your Study Plan

The answer key on the following page shows where in this chapter to find explanations for the questions you missed. Here's how you should proceed based on your Diagnostic Test score.

5: Superb! You have a solid grasp of the essentials of solid geometry, and if you're pressed for time, you might consider skipping this chapter. If you want to try your hand at a few more solid geometry questions, go straight to the Follow-Up Test at the end of this chapter.

3–4: Good! If you can get three or four of these relatively difficult questions right, then you have a decent understanding of solid geometry. If you're pressed for time, you might consider skipping this chapter. You might want to look over the parts of this chapter that deal with the one or two questions you didn't get right. If you just want to try your hand at a few more solid geometry questions, go straight to the Follow-Up Test at the end of this chapter.

0–2: Solid geometry is not your forte right now, but since it accounts for only three or four questions on the test, it may not be worth worrying about. If you're pressed for time, and are confident with the other test topics, you might consider skipping this chapter. But if you have the time, you should read the rest of this chapter, study the examples, and then try the Follow-Up Test at the end of the chapter.

Solid Geometry Test Topics

The questions in the Solid Geometry Diagnostic Test are typical of the SAT II: Mathematics Subject Tests. In this chapter we'll use these questions to review the solid geometry you're expected to know for the Mathematics Subject Tests. We will also use these questions to demonstrate some effective problem-solving techniques, alternative methods, and test-taking strategies that apply to SAT II solid geometry questions.

 Five Formulas You Don't Need to Memorize

Some people find solid geometry intimidating because of all those formulas. On the SAT II, however, some of the scariest formulas are given to you in the directions. It's good to know what formulas are included in the directions so you don't waste time memorizing them, and so that you won't forget to memorize all the relevant formulas that are *not* included in the directions. These are the formulas that are printed in the directions.

Test Topics

This icon appears next to each discussion of a math topic that's tested on the SAT II.

Diagnostic Test Answers and Reviews

1. E
See "Five Formulas You Don't Need to Memorize," p. 101.

2. C
See "The Test Makers' Favorite Solid," p. 105.

3. E
See "Uniform Solids," p. 107.

4. E
See "Picturing Solids— Rotating Polygons," p. 110.

5. A
See "Picturing Solids— Maximums and Counting," p. 111.

Lateral area of a cone: Given base circumference c and slant height ℓ,

$$\text{Lateral Area of Cone} = \frac{1}{2}c\ell$$

The lateral area of a cone is the area of the part that extends from the vertex to the circular base. It does not include the circular base.

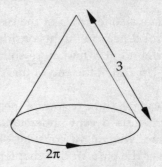

For example, in the figure above, $c = 2\pi$ and $\ell = 3$, so:

$$\text{Lateral Area} = \frac{1}{2}(2\pi)(3) = 3\pi$$

Volume of a cone: Given base radius r and height h,

$$\text{Volume of Cone} = \frac{1}{3}\pi r^2 h$$

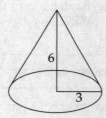

For example, in the figure above, $r = 3$, and $h = 6$, so:

$$\text{Volume} = \frac{1}{3}\pi\left(3^2\right)(6) = 18\pi$$

Surface area of a sphere: Given radius r,

$$\text{Surface Area of Sphere} = 4\pi r^2$$

For example, if the radius of a sphere is 2, then:

$$\text{Surface Area} = 4\pi(2^2) = 16\pi$$

Volume of a sphere: Given radius r,

$$\text{Volume of Sphere} = \frac{4}{3}\pi r^3$$

For example, if the radius of a sphere is 2, then:

$$\text{Volume} = \frac{4}{3}\pi(2)^3 = \frac{32\pi}{3}$$

Volume of a pyramid: Given base area B and height h,

$$\text{Volume of Pyramid} = \frac{1}{3}Bh$$

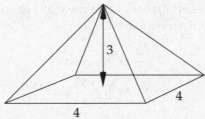

For example, in the figure above, $h = 3$ and, if the base is a square, $B = 16$, then:

$$\text{Volume} = \frac{1}{3}(16)(3) = 16$$

Here's a question that uses some of these formulas.

Example 1

A right circular cone and a sphere have equal volumes. If the radius of the base of the cone is $2x$ and the radius of the sphere is $3x$, what is the height of the cone in terms of x?

- (A) x

- (B) $\dfrac{3x}{2}$

- (C) $\dfrac{4x}{3}$

- (D) $20x$

- (E) $27x$

As you can see, this question is hard enough even with the formulas provided. This is no mere matter of plugging values into a formula and cranking out the answer. This question is more algebraic than that and takes a little thought. It's really a word problem. It describes in words a mathematical situation (in this case, geometric) that can be translated into algebra. The pivot in this situation is that the cone and sphere have equal

Five Formulas You Don't Need to Memorize

These solids formulas are included in the SAT II: Mathematics Test directions.

Lateral Area of a Cone = $\dfrac{1}{2}c\ell$

Volume of a Cone = $\dfrac{1}{3}\pi r^2 h$

Surface Area of a Sphere = $4\pi r^2$

Volume of a Sphere = $\dfrac{4}{3}\pi r^3$

Volume of a Pyramid = $\dfrac{1}{3}Bh$

volumes. You're looking for the height h in terms of x, and fortunately you can express both volumes in terms of those two variables. Be careful. Both formulas include r, but they're not the same r's. In the case of the cone, $r = 2x$, but in the case of the sphere, $r = 3x$:

$$\text{Volume of cone} = \frac{1}{3}\pi r^2 h = \frac{1}{3}\pi(2x)^2 h = \frac{4}{3}\pi x^2 h$$

$$\text{Volume of sphere} = \frac{4}{3}\pi r^3 = \frac{4}{3}\pi(3x)^3 = 36\pi x^3$$

Now write an equation that says that the expressions for the two volumes are equal to each other, and solve for h :

$$\frac{4}{3}\pi x^2 h = 36\pi x^3$$
$$\pi x^2 h = \frac{3}{4}\left(36\pi x^3\right)$$
$$\pi x^2 h = 27\pi x^3$$
$$h = \frac{27\pi x^3}{\pi x^2} = 27x$$

And so the answer is (E).

So the direct way to do this question is the algebraic way. But as you can see, the algebra is quite convoluted, and the matter of the different r's can be confusing. There's another, less sophisticated way to do this question— *pick numbers*. All the given measures are in terms of x, and all the answer choices are in terms of x, so to make things simpler, you could just pick a number for x, plug it into the question, and see what you get. Pick a number that's easy to work with. Here you could even pick $x = 1$ because it's clear that when $x = 1$, all the answer choices have different values.

If $x = 1$, then the radius of the base of the cone is 2 and the radius of the sphere is 3.

$$\text{Volume of cone} = \frac{1}{3}\pi r^2 h = \frac{1}{3}\pi(2)^2 h = \frac{4}{3}\pi h$$

$$\text{Volume of sphere} = \frac{4}{3}\pi r^3 = \frac{4}{3}\pi(3)^3 = 36\pi$$

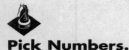

Pick Numbers.

When the answer choices are algebraic expressions, it often works to pick numbers for the unknowns, plug those numbers into the stem and see what you get, and then plug those same numbers into the answer choices to find matches.

Warning: When you pick numbers, you have to check all the answer choices. Sometimes more than one works with the number(s) you pick, in which case you have to pick numbers again.

So what does h have to be to give the cone a volume of 36π ?

$$\frac{4}{3}\pi h = 36\pi$$

$$\frac{4}{3}h = 36$$

$$h = 36 \times \frac{3}{4} = 27$$

When $x = 1$, h turns out to be 27. Now plug $x = 1$ into the answer choices and you'll find that only (E) gives you 27. This alternative method is still somewhat algebraic, but this way the algebra's a lot less convoluted.

The Test Makers' Favorite Solid

The test makers' favorite solid is the *rectangular solid*. That's the official geometric term for a box, which has six rectangular faces, twelve edges that meet at right angles at eight vertices.

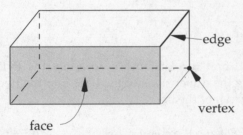

The *surface area* of a rectangular solid is simply the sum of the areas of the faces. That's what the formula "Surface Area = $2\ell w + 2wh + 2\ell h$" says. If the length is ℓ, the width is w, and the height is h, then two rectangular faces have area ℓw, two have area wh, and two have area ℓh. The total surface area is the sum of those three pairs of areas.

Instead of the surface area, you may be asked to find the distance between opposite vertices of a rectangular solid. Look at Example 2.

Brush Up on Your Algebra.

You can't escape algebra! The SAT II: Mathematics Subject Tests are both full of algebra. Even a lot of the geometry questions are partly algebra questions as well. Make sure your algebra skills are in peak condition by Test Day.

Surface Area of a Rectangular Solid

If the length is ℓ, the width is w, and the height is h, the formula is:
Surface Area = $2\ell w + 2wh + 2\ell h$

Example 2

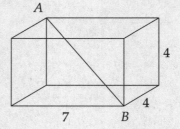

Figure 1

In the rectangular solid in Figure 1, what is the distance from vertex A to vertex B ?

(A) $\sqrt{65}$

(B) $7\sqrt{2}$

(C)   9

(D)   11

(E)   15

One way to find this distance is to apply the Pythagorean theorem twice. First plug the dimensions of the base into the Pythagorean theorem to find the diagonal of the base:

$$\text{Diagonal of base} = \sqrt{4^2 + 7^2} = \sqrt{16 + 49} = \sqrt{65}$$

Notice that the base diagonal combines with an edge and with the segment AB you're looking for to form a right triangle:

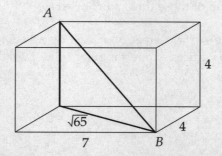

So you can plug the base diagonal and the height into the Pythagorean theorem to find AB:

$$AB = \sqrt{\left(\sqrt{65}\right)^2 + 4^2} = \sqrt{65 + 16} = \sqrt{81} = 9$$

The answer is (C).

Another way to find this distance is to use the formula, which you could say is just the Pythagorean theorem taken to another dimension. If the length is ℓ, the width is w, and the height is h, the formula is:

$$\text{Distance} = \sqrt{\ell^2 + w^2 + h^2}$$

Distance Between Opposite Vertices of a Rectangular Solid

$$\text{Distance} = \sqrt{\ell^2 + w^2 + h^2}$$

 ## Uniform Solids

A rectangular solid is one type of *uniform solid*. A uniform solid is what you get when you take a plane and move it, without tilting it, through space. Here are some uniform solids.

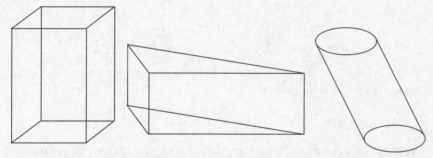

The way these solids are drawn, the top and bottom faces are parallel and congruent. These faces are called the *bases*. You can think of each of these solids as the result of sliding the base through space. The perpendicular distance through which the base slides is called the *height*.

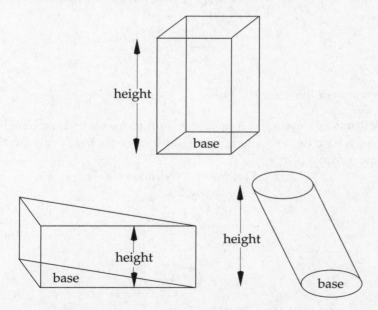

Volume Formulas for Uniform Solids

Volume of a Uniform Solid = Bh

Volume of a Rectangular Solid = ℓwh

Volume of a Cube = e^3

Volume of a Cylinder = $\pi r^2 h$

In every one of the above cases—indeed, in the case of *any* uniform solid—the volume is equal to the area of the base times the height. So, you can say that for any uniform solid, given the area of the base B and the height h,

Volume of a Uniform Solid = Bh

Volume of a rectangular solid: A rectangular solid is a uniform solid whose base is a rectangle. Given the length ℓ, width w, and height h, the area of the base is ℓw, and so the volume formula is:

Volume of a Rectangular Solid = ℓwh

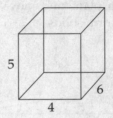

The volume of a 4-by-5-by-6 box is

$$4 \times 5 \times 6 = 120$$

Volume of a cube: A cube is a rectangular solid with length, width, and height all equal. If e is the length of an edge of a cube, the volume formula is:

Volume of a Cube = e^3

The volume of this cube is $2^3 = 8$.

Volume of a cylinder: A *cylinder* is a uniform solid whose base is a circle. Given base radius r and height h, the area of the base is πr^2, and so the volume formula is:

Volume of a Cylinder = $\pi r^2 h$

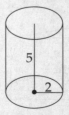

In the cylinder above, $r = 2$ and $h = 5$, so:

$$\text{Volume} = \pi(2^2)(5) = 20\pi$$

Example 3 gives you the volume of a uniform solid and asks for the surface area.

Example 3

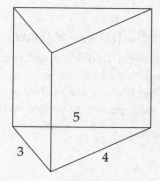

Figure 2

In Figure 2, the bases of the right uniform solid are triangles with sides of lengths 3, 4, and 5. If the volume of the solid is 30, what is the total surface area?

(A)   17

(B)   30

(C)   36

(D)   60

(E)   72

The surface area is the sum of the areas of the faces. To find the areas of the faces, you need to figure out what kinds of polygons they are so that you'll know what formulas to use. Start with the bases, which are said to be "triangles with sides of lengths 3, 4, and 5." If side lengths don't ring a bell in your head, then you'd better go back to Chapter 5 and bone up on your special triangles. This is a 3-4-5 triangle, which means that it's a right triangle, which means that you can use the legs as the base and height to find the area:

$$\text{Area of Right Triangle} = \frac{1}{2}(\text{leg}_1)(\text{leg}_2) = \frac{1}{2}(3)(4) = 6$$

That's the area of each of the bases. The other three faces are rectangles. To find their areas, you need first to determine the height of the solid. If the area of the base is 6, and the volume is 30, then:

$$Volume = Bh$$
$$30 = 6h$$
$$h = 5$$

So the areas of the three rectangular faces are $3 \times 5 = 15$, $4 \times 5 = 20$, and $5 \times 5 = 25$. The total surface area, then, is $6 + 6 + 15 + 20 + 25 = 72$. The answer is (E).

 ## Picturing Solids—Rotating Polygons

You might have thought that solid geometry questions are difficult because of the formulas. As you have seen, however, using formulas is by definition routine. It's the nonroutine problems that can be the most challenging. Those are the solid geometry problems that require you to visualize, like Example 4:

Example 4

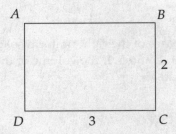

Figure 3

If the rectangle in Figure 3 is rotated 360° about side BC, what is the volume of the resulting solid?

(A)   12.00

(B)   18.00

(C)   28.27

(D)   37.70

(E)   56.55

Can you visualize—picture in your mind—what the resulting solid looks like? It can be a little difficult to sketch 3-D geometry. You're probably better off if you can just "see" it in your head. It'll look something like this.

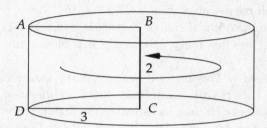

It's a cylinder with base radius 3 and height 2. So now all you have to do is plug $r = 3$ and $h = 2$ into the formula:

$$\text{Volume of Cylinder} = \pi r^2 h = \pi(3^2)(2) = 18\pi \approx 56.55$$

The answer is (E). Applying the formula was the easy part. It was visualizing the cylinder and figuring out what's r and what's h that was the challenging part. A typical SAT II: Mathematics Subject Test will include one or two questions that entail visualizing.

Picturing Solids—Maximums and Counting

Many questions that are more about visualizing than about applying formulas are questions that ask you to find a maximum or to count possibilities. In Example 5 you are asked to count up the maximum number of blocks that will fit inside a carton.

Example 5

What is the maximum number of rectangular blocks, each with dimensions 3 centimeters by 5 centimeters by 7 centimeters, that can be placed inside a rectangular carton with inside dimensions 14 centimeters by 15 centimeters by 16 centimeters ?

(A)   30

(B)   32

(C)   33

(D)   35

(E)   45

If you try to do this question with formulas and without visualizing, you'll probably get the wrong answer. You might think that all you have to do is find the volume of the inside of the carton and the volume of one block and divide. The volume of the inside of the carton is $14 \times 15 \times 16 = 3{,}360$. The volume of one block is $3 \times 5 \times 7 = 105$. When you divide, you get 32. So the volume of the inside of the carton is exactly 32 times the volume of one block. Does that mean you can pack 32 blocks into the carton?

Not necessarily. That method assumes that you can fit the blocks exactly inside the carton with no unfilled space. In this case, however, that's impossible. You can fit the blocks inside the carton with no unfilled space only if you can match each of the 3 dimensions of the carton with a different dimension of a block such that the smaller number divides into the larger with no remainder.

Here you can match the 7 with the 14 and the 3 with the 15, but then what's left is 5 and 16, which would leave a remainder of 1. Or you could match the 7 with the 14 and the 5 with the 15, but then what's left is 3 and 16, which would also leave a remainder of 1. Any way you try to pack the carton, you're going to have some unfilled space. Can you visualize it?

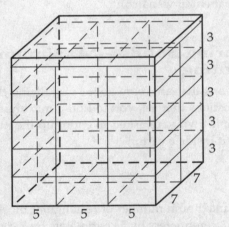

The 5 edge fits into the 15 edge exactly 3 times. The 7 edge fits into the 14 edge exactly 2 times. And the 3 edge fits into the 16 edge 5 times with some left over. The total number of blocks in this configuration is $2 \times 3 \times 5 = 30$. There's no way to get more blocks in there, so the answer is (A).

You've covered the solid geometry topics that you're likely to encounter on the SAT II: Mathematics Tests. You've reviewed the facts and formulas, and learned some useful strategies. Now it's time for you to try another set of typical SAT II solid geometry questions.

Solid Geometry Follow-Up Test

5 Questions (6 Minutes)

Directions: Solve the following problems and choose the best answer from those given. Fill in the oval corresponding to the best answer choice in the grid to the right of each question. (Answers and explanations begin on page 116.)

Reference Information: Use the following formulas as needed.

Right circular cone: If r = radius and h = height, then **Volume** $= \frac{1}{3}\pi r^2 h$; and if c = circumference of the base and ℓ = slant height, then **Lateral Area** $= \frac{1}{2}c\ell$.

Sphere: If r = radius, then **Volume** $= \frac{4}{3}\pi r^3$ and **Surface Area** $= 4\pi r^2$.

Pyramid: If B = area of the base and h = height, then **Volume** $= \frac{1}{3}Bh$.

DO YOUR FIGURING HERE.

1. In Figure 1, the radius of the base of the right circular cone is 3. If the volume of the cone is 12π, what is the lateral area of the cone?

    (A) 12π

    (B) 15π

    (C) 18π

    (D) 36π

    (E) 72π

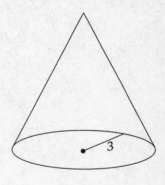

Figure 1

DO YOUR FIGURING HERE.

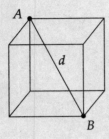

Figure 2

2. In Figure 2, d is the distance from vertex A to vertex B. What is the volume of the cube in terms of d ?

    (A) $9d^3\sqrt{3}$

    (B) $3d^3\sqrt{3}$

    (C) $d^3\sqrt{3}$

    (D) $\dfrac{d^3\sqrt{3}}{3}$

    (E) $\dfrac{d^3\sqrt{3}}{9}$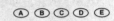

3. A cube with edge of length 4 is divided into 8 identical cubes. How much greater is the combined surface area of the 8 smaller cubes than the surface area of the original cube?

    (A) 48

    (B) 56

    (C) 96

    (D) 288

    (E) 384

4. When a right triangle of area 3 is rotated 360° about its shorter leg, the solid that results has a volume of 30. What is the volume of the solid that results when the same right triangle is rotated about its longer leg?

    (A) 0.99

    (B) 7.90

    (C) 8.88

    (D) 31.42

    (E) 41.12

5. The pyramid in Figure 3 is composed of a square base of area 12 and four equilateral triangles. What is the volume of the pyramid?

   (A)   9.8

   (B)   14.7

   (C)   17.0

   (D)   19.6

   (E)   29.4

DO YOUR FIGURING HERE.

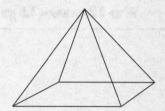

Figure 3

ⒶⒷⒸⒹⒺ

STOP!   END OF TEST. DO NOT TURN THE PAGE UNTIL YOU ARE READY TO CHECK YOUR ANSWERS.

Follow-Up Test—Answers and Explanations

1. (B)—The formula for the lateral surface area of a cone is in terms of c = circumference and ℓ = slant height. You can use the given base radius to get the circumference:

$$c = 2\pi r = 2\pi(3) = 6\pi$$

The slant height you can think of as the hypotenuse of a right triangle whose legs are the base radius and height of the cone.

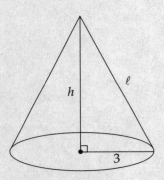

To get ℓ, first you need to find h :

$$\text{Volume of Cone} = \frac{1}{3}\pi r^2 h$$

$$12\pi = \frac{1}{3}\pi(3^2)h$$

$$12\pi = 3\pi h$$

$$h = \frac{12\pi}{3\pi} = 4$$

Now you can see that the triangle is a 3-4-5 and that $\ell = 5$. Now plug $c = 6\pi$ and $\ell = 5$ into the lateral area formula:

$$\text{Lateral Area} = \frac{1}{2}c\ell = \frac{1}{2}(6\pi)(5) = 15\pi$$

2. (E)—Use the formula for the distance between opposite vertices:

$$d = \sqrt{\ell^2 + w^2 + h^2}$$

What you have is a cube, so the length, width, and height are all the same—call them each x:

$$d = \sqrt{x^2 + x^2 + x^2} = \sqrt{3x^2} = x\sqrt{3}$$

$$x = \frac{d}{\sqrt{3}}$$

Now you have the length of an edge in terms of d. Cube that and you have the volume in terms of d :

$$\text{Volume} = (\text{edge})^3$$

$$= x^3$$

$$= \left(\frac{d}{\sqrt{3}}\right)^3$$

$$= \frac{d^3}{3\sqrt{3}} \cdot \frac{\sqrt{3}}{\sqrt{3}}$$

$$= \frac{d^3\sqrt{3}}{9}$$

3. (C)—When a cube of edge length 4 is divided into 8 identical smaller cubes, the edge of each of the smaller cubes is 2:

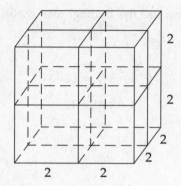

The surface area of the original cube is $6 \times 4 \times 4 = 96$. The surface area of one of the smaller cubes is $6 \times 2 \times 2 = 24$. There are eight small cubes, so their combined surface area is $8 \times 24 = 192$. The difference is $192 - 96 = 96$.

4. **(B)**—When you rotate a right triangle about a leg, you get a cone:

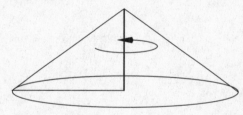

One leg becomes the base radius and the other becomes the height of the cone. Call the long leg a and the short leg b and plug them into the cone volume formula:

$$\text{Volume of Cone} = \frac{1}{3}\pi r^2 h = \frac{1}{3}\pi a^2 b$$

It's also given that the area of the right triangle is 3, so:

$$\text{Area of Right Triangle} = \frac{1}{2}\big(\text{leg}_1\big)\big(\text{leg}_2\big)$$
$$3 = \frac{1}{2}ab$$
$$ab = 6$$

Plug $ab = 6$ into the expression for the volume of the cone and you can solve for a:

$$\frac{1}{3}\pi a^2 b = 30$$
$$\frac{1}{3}\pi a(ab) = 30$$
$$\frac{1}{3}\pi a(6) = 30$$
$$2\pi a = 30$$
$$a = \frac{15}{\pi}$$

Then you can plug $a = \dfrac{15}{\pi}$ into the equation $ab = 6$ to solve for b:

$$ab = 6$$
$$\left(\frac{15}{\pi}\right)b = 6$$
$$b = \frac{6\pi}{15} = \frac{2\pi}{5}$$

When the triangle is rotated about its longer leg, a becomes the height and b becomes the base radius, so:

$$\text{Volume} = \frac{1}{3}\pi r^2 h = \frac{1}{3}\pi b^2 a = \frac{1}{3}\pi \left(\frac{2\pi}{5}\right)^2\left(\frac{15}{\pi}\right) = \frac{60\pi^3}{75\pi}$$
$$= \frac{4}{5}\pi^2 \approx 7.90$$

5. (A)—To find the volume of a pyramid, you need the area of the base, which is given here as 12, and you need the height, which here you have to figure out. Imagine a triangle that includes the height and one of the lateral edges:

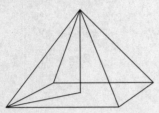

This triangle is a right triangle. The hypotenuse is the same as a side of one of the equilateral triangles, which is the same as a side of the square, which is the square root of 12, or $2\sqrt{3}$:

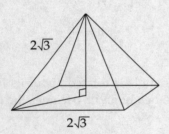

One of the legs of this right triangle is half of a diagonal of the square base—that is, half of $\left(2\sqrt{3}\right)\sqrt{2}$:

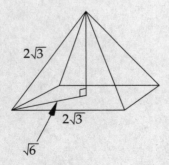

Now you can use the Pythagorean theorem to find the height:

$$\left(\sqrt{6}\right)^2 + h^2 = \left(2\sqrt{3}\right)^2$$
$$6 + h^2 = 12$$
$$h^2 = 6$$
$$h = \sqrt{6}$$

Finally you have what you need to use the volume formula:

$$\text{Volume} = \frac{1}{3}Bh = \frac{1}{3}(12)\sqrt{6} = 4\sqrt{6} \approx 9.8$$

What's Next

No matter what level SAT II: Mathematics Test you're planning to take, you'll need the information about coordinate geometry in the next chapter.

CHAPTER 7

Coordinate Geometry

A typical SAT II: Mathematics Subject Test—whether it's Level IC or Level IIC—has about six coordinate geometry questions. But that's counting just the questions that are primarily about coordinate geometry. A lot of trigonometry and functions questions assume an understanding of coordinate geometry. In fact, the material in this chapter is relevant to approximately *20 percent of Level IC questions and 30 percent of IIC questions.*

How to Use This Chapter

Maybe you already know all the coordinate geometry you need. You can find out by taking the Coordinate Geometry Diagnostic Test on page 123. Five of the six coordinate geometry questions on the Diagnostic Test are typical of what you could expect on either a Level IC or a Level IIC Mathematics Subject Test. One question on the Diagnostic, though, could appear only on a Level IIC Test. We've marked that Level IIC question with a star. Check your answers using the answer key following the test. No matter how you score, don't worry! The answer key also shows where to find a detailed explanation for each question. The "Find Your Study Plan" section that follows the test will suggest next steps based on your performance on the Diagnostic.

 Find Your Level

How you use this chapter really depends on which test you're taking and how much time you have to prep. Find your level and pace below.

Taking the Mathematics Level IC Test? Answer the first five questions in the Coordinate Geometry Diagnostic Test. Skip the Level IIC question that we've marked with a star. Read the rest of the chapter, skipping any material that we've flagged as relating to Level IIC only. Then try the Follow-Up Test at the end.

Coordinate Geometry Facts and Formulas in This Chapter

- Midpoint (p. 127)
- Distance Formula (p. 127)
- Slope-Intercept Equation Form (p. 130)
- Slope Formula (p. 130)
- Positive and Negative Slopes (p. 131)
- Slopes of Parallel and Perpendicular Lines (p. 132)
- Circle (p. 135)
- Parabola (p. 135)

Topics Relating to Level IIC only:

- Ellipse (p. 136)★
- Hyperbola (p. 136)★

★
What's This Mean?

This icon appears next to more difficult topics and questions that would appear only on the Level IIC Test.

Kaplan Strategies in This Chapter

- Picture it (p. 128).
- Don't just memorize— internalize (p. 131).
- Pick a point (p. 134).

Mathematics Level IC Shortcut Answer the first five questions in the Coordinate Geometry Diagnostic Test. Skip the Level IIC question that we've marked with a star. The "Find Your Study Plan" section that follows the test will suggest next steps based on your Diagnostic Test score.

Taking the Mathematics Level IIC Test? Do everything in this chapter. It's all relevant to the Level IIC Test.

Mathematics Level IIC Shortcut Take the Coordinate Geometry Diagnostic Test and check your answers. If you can answer most of the questions correctly, you should skip this chapter.

Panic Plan? Look through the chapter quickly, and make sure you're comfortable with the material. If you're not comfortable with coordinate geometry, you should probably spend at least a little time with this chapter before moving on.

Coordinate Geometry Diagnostic Test

Level IC: 5 Questions (6 Minutes)
Level IIC: 6 Questions (8 Minutes)

<u>Directions for Level IC</u>: If you are preparing for the SAT II: Mathematics Level IC Test, solve problems 1–5 and choose the best answer from those given. Fill in the oval corresponding to the best answer choice in the grid to the right of each question. (Answers are on page 126.)

<u>Directions for Level IIC</u>: If you are preparing for the SAT II: Mathematics Level IIC Test, solve problems 1–6 and choose the best answer from those given. Fill in the oval corresponding to the best answer choice in the grid to the right of each question. (Answers are on page 126.)

1. In Figure 1, what is the distance from the midpoint of segment *AC* to the midpoint of segment *BD* ?

    (A)   1.118

    (B)   1.414

    (C)   1.803

    (D)   2.236

    (E)   2.828   Ⓐ Ⓑ Ⓒ Ⓓ Ⓔ

2. If points (6, 0), (0, 0), (0, 2), and (*a*, 2) are consecutive vertices of a trapezoid of area 7.5, what is the value of *a* ?

    (A)   1.5

    (B)   2

    (C)   2.5

    (D)   5

    (E)   9   Ⓐ Ⓑ Ⓒ Ⓓ Ⓔ

DO YOUR FIGURING HERE.

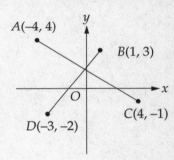

Figure 1

DO YOUR FIGURING HERE.

3. Which of the following lines has no point of intersection with the line $y = 4x + 5$?

(A) $y = \dfrac{1}{4}x - 5$

(B) $y = -\dfrac{1}{4}x - 5$

(C) $y = 4x + \dfrac{1}{5}$

(D) $y = -4x + \dfrac{1}{5}$

(E) $y = -4x - \dfrac{1}{5}$

Ⓐ Ⓑ Ⓒ Ⓓ Ⓔ

4. Which of the following shaded regions shows the graph of the inequality $y \le |x + 2|$?

(A)
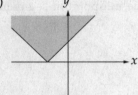

(B)
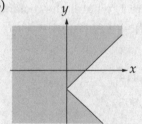

(C)
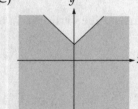

(D)
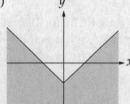

(E)
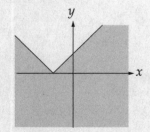

Ⓐ Ⓑ Ⓒ Ⓓ Ⓔ

5. Which of the following equations describes the set of all points (x, y) in the coordinate plane that are a distance of 5 from the point $(-3, 4)$?

    (A) $(x + 3) + (y - 4) = 5$

    (B) $(x - 3) + (y + 4) = 5$

    (C) $(x + 3)^2 + (y - 4)^2 = 5$

    (D) $(x + 3)^2 + (y - 4)^2 = 25$

    (E) $(x - 3)^2 + (y + 4)^2 = 25$ Ⓐ Ⓑ Ⓒ Ⓓ Ⓔ

★6. Which of the following is an equation of an ellipse centered at the origin and with axial intersections at $(0, \pm 3)$ and $(\pm 2, 0)$?

    (A) $\dfrac{x}{2} + \dfrac{y}{3} = 1$

    (B) $\dfrac{x}{2} + \dfrac{y}{3} = 2$

    (C) $\dfrac{x}{3} + \dfrac{y}{2} = 2$

    (D) $\dfrac{x^2}{4} + \dfrac{y^2}{9} = 1$

    (E) $\dfrac{x^2}{4} + \dfrac{y^2}{9} = 2$ Ⓐ Ⓑ Ⓒ Ⓓ Ⓔ

DO YOUR FIGURING HERE.

STOP! END OF TEST. DO NOT TURN THE PAGE UNTIL YOU ARE READY TO CHECK YOUR ANSWERS.

Diagnostic Test Answers and Reviews

1. B
See "Midpoints and Distances," p. 126.

2. A
See "Geometry on the Grid," p. 128.

3. C
See "Slope-Intercept Equation Form," p. 130.

4. E
See "Absolute Value and Inequalities," p. 133.

5. D
See "Circles and Parabolas," p. 134.

6. D
See "Ellipses and Hyperbolas," p. 135.★

Test Topics

This icon appears next to each discussion of a math topic that's tested on the SAT II.

Find Your Study Plan

The answer key shows where in this chapter to find explanations for the questions you missed. Here's how you should proceed based on your Diagnostic Test score.

5–6: Superb! If there's not much time before Test Day, then you might consider skipping this chapter—you seem to know it all already! If you just want to try your hand at some more coordinate geometry questions, go to the Follow-Up Test at the end of this chapter.

3–4: Good. You have a decent grasp of coordinate geometry. But this topic is fundamental enough that you might want to read at least those parts of this chapter that relate to the questions you were unable to answer correctly.

0–2: You need to work on coordinate geometry. This topic is fundamental. Read the rest of this chapter, study the examples, and see if you can do better on the Follow-Up Test at the end of the chapter.

Coordinate Geometry Test Topics

The questions in the Coordinate Geometry Diagnostic Test are typical of the SAT II: Mathematics Test. In this chapter we'll use these questions to review the coordinate geometry you're expected to know for both levels of the Mathematics Subject Test. We will also use these questions to demonstrate some effective problem-solving techniques, alternative methods, and test-taking strategies that apply to SAT II coordinate geometry questions.

Midpoints and Distances

Some of the more basic SAT II coordinate geometry questions are ones that concern themselves with the layout of the grid, the location of points, distances between them, midpoints, and so on. Example 1 involves both distances and midpoints.

Example 1

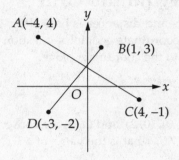

Figure 1

In Figure 1, what is the distance from the midpoint of segment AC to the midpoint of segment BD ?

(A)   1.118

(B)   1.414

(C)   1.803

(D)   2.236

(E)   2.828

To find the midpoint of a segment, average the x-coordinates and average the y-coordinates of the endpoints:

$$\text{midpoint of } AC = \left(\frac{-4+4}{2}, \ \frac{4-1}{2} \right) = (0, \ 1.5)$$

$$\text{midpoint of } BD = \left(\frac{-3+1}{2}, \ \frac{-2+3}{2} \right) = (-1, \ 0.5)$$

To find the distance between two points, use the distance formula:

$$\text{Distance} = \sqrt{(x_2 - x_1)^2 + (y_2 - y_1)^2}$$

The distance from $(0, 1.5)$ to $(-1, 0.5)$ is:

$$\begin{aligned} \text{Distance} &= \sqrt{(-1-0)^2 + (0.5-1.5)^2} \\ &= \sqrt{1+1} \\ &= \sqrt{2} \end{aligned}$$

You could use your calculator to get a decimal approximation of $\sqrt{2}$, but you should be able to spot it as (B) 1.414. The answer is (B).

Midpoint

To find the midpoint between (x_1, y_1) and (x_2, y_2), average the x-coordinates and average the y-coordinates:

$$\text{Midpoint} = \left(\frac{x_1 + x_2}{2}, \ \frac{y_1 + y_2}{2} \right)$$

Distance Formula

To find the distance between (x_1, y_1) and (x_2, y_2):

Distance =

$$\sqrt{(x_2 - x_1)^2 + (y_2 - y_1)^2}$$

Geometry on the Grid

Finding midpoints and distances is basically a geometry thing. Here's another example of a coordinate geometry question that's essentially plane geometry transferred to the coordinate plane.

Example 2

If points $(6, 0)$, $(0, 0)$, $(0, 2)$, and $(a, 2)$ are consecutive vertices of a trapezoid of area 7.5, what is the value of a ?

(A)   1.5

(B)   2

(C)   2.5

(D)   5

(E)   9

Picture It.

Whenever you come to a geometry question with no figure, you should normally sketch one.

Unless you're really good at visualizing, you should sketch a diagram to help you comprehend the situation. Plot the three given points and connect them to make two of the sides of the trapezoid:

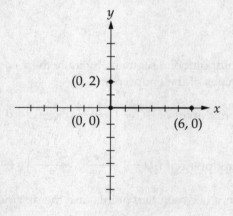

The fourth vertex has a y-coordinate of 2, so it must be somewhere along the line $y = 2$:

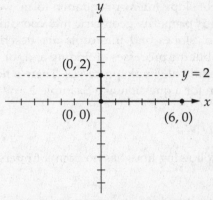

That will make the top and bottom sides of the trapezoid the parallel bases. The formula for the area of a trapezoid is:

$$\text{Area of Trapezoid} = \left(\frac{\text{base}_1 + \text{base}_2}{2}\right) \times \text{height}$$

Here it's given that the area is 7.5, and you can see from the figure that one base is 6 and the height is 2. That's enough to solve for the other base.

$$7.5 = \left(\frac{6 + \text{base}_2}{2}\right) \times 2$$

$$\text{base}_2 = 7.5 - 6 = 1.5$$

The top base is 1.5, and so that the four vertices will be consecutive, the coordinates of the fourth vertex are (1.5, 2).

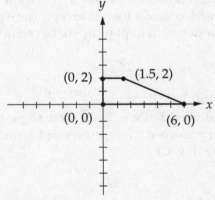

So $a = 1.5$, and the answer is (A).

Slope-Intercept Equation Form

For any equation in the form
$y = mx + b$:
  m = slope
  b = y-intercept

 ### Slope-Intercept Equation Form

With the topic of slope-intercept equation form, we move from coordinate geometry that is primarily geometric into coordinate geometry that is primarily algebraic. Slopes and intercepts are descriptions of lines and points on the grid, but the processes of finding and/or using slopes and/or intercepts are generally algebraic processes. There's no need, for instance, to sketch a diagram for a question like Example 3.

Example 3

Which of the following lines has no point of intersection with the line $y = 4x + 5$?

(A) $y = \dfrac{1}{4}x - 5$

(B) $y = -\dfrac{1}{4}x - 5$

(C) $y = 4x + \dfrac{1}{5}$

(D) $y = -4x + \dfrac{1}{5}$

(E) $y = -4x - \dfrac{1}{5}$

What does *has no intersection with* mean? It means that the lines are parallel, which in turn means that the lines have the same slope. If you know the slope-intercept equation form, you're able to spot the correct answer instantly.

Slope Formula

Slope is defined as
$\dfrac{\text{change in } y}{\text{change in } x}$.

To find the slope of a line containing the points (x_1, y_1) and (x_2, y_2):

Slope = $\dfrac{y_2 - y_1}{x_2 - x_1}$

When an equation is in the form $y = mx + b$, the letter m represents the slope and the letter b represents the y-intercept. The equation in the stem is $y = 4x + 5$. That's in slope-intercept form, so the coefficient of x is the slope.

$$y = \textcircled{4}x + 5$$
$$\searrow \text{slope} = 4$$

Now look for the answer choice with the same slope. Conveniently, all the answer choices are presented in slope-intercept form, so spotting the one with $m = 4$ is a snap. It's (C):

$$y = \textcircled{4}x + \dfrac{1}{5}$$
$$\searrow \text{slope} = 4$$

People who are good at memorizing methods and formulas are not necessarily the ones who get the best scores on the Mathematics Subject Tests. It's people who have a deeper understanding of mathematics. If you really want to ace coordinate geometry questions, it's not enough to memorize the midpoint formula, the distance formula, the slope definition, the slope-intercept equation form, and so on. What you want is to have a real grasp of what *slope* is, what *perpendicular, parallel, positive, negative, zero,* and *undefined* slopes tell you.

Slope is a description of the "steepness" of a line. Lines that go uphill (from left to right) have positive slopes:

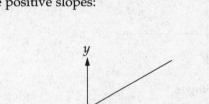

Positive Slopes

Lines that go downhill have negative slopes:

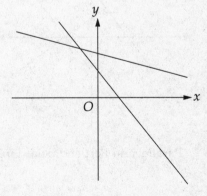

Negative Slopes

Don't Just Memorize— Internalize.

The best scores go to the test takers who don't just memorize methods and formulas but who really understand the underlying math.

Positive and Negative Slopes

Lines that go uphill from left to right have positive slopes. The steeper the uphill grade, the greater the slope.

Lines that go downhill from left to right have negative slopes. The steeper the downhill grade, the less the slope.

Slopes of Parallel and Perpendicular Lines

Parallel lines have the same slope.

Perpendicular lines have negative-reciprocal slopes.

Lines parallel to the *x*-axis have slope = 0, and lines parallel to the *y*-axis have undefined slope.

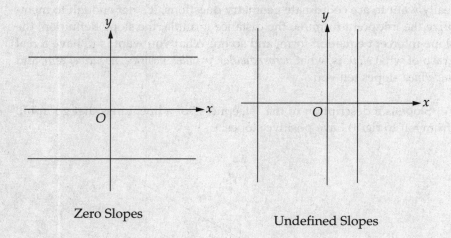

Zero Slopes   Undefined Slopes

Lines that are parallel to each other have the same slope, and lines that are perpendicular to each other have negative-reciprocal slopes. In the figure below, the two parallel lines both have slope = 2, and the line that's perpendicular to them has slope = $-\dfrac{1}{2}$:

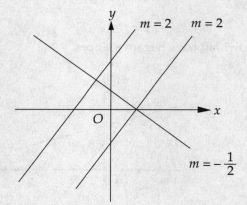

Parallel and Perpendicular Lines

Absolute Value and Inequalities

Among the more mind-bending coordinate geometry questions you'll find in the Level IC test are ones like Example 4, which entail graphing absolute value and inequalities.

Example 4

Which of the following shaded regions shows the graph of the inequality $y \le |x + 2|$?

(A)

(B)
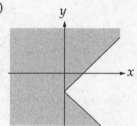

(C)
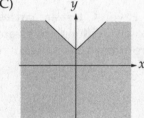

(D)
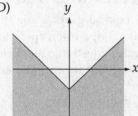

(E)
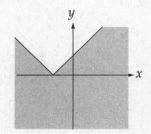

133

The way to handle an inequality is to think of it as an equation first, plot the line, and then figure out which side of the line to shade. This inequality is extra complicated because of the absolute value. When you graph an equation with absolute value, you generally get a line with a bend, as in all of the answer choices above. To find the graph of an absolute value equation, figure out where that bend is. In this case $|x + 2|$ has a turning-point value of 0, which happens when $x = -2$. So the bend is at the point $(-2, 0)$. That narrows the choices down to (A) and (E).

Next, figure out which side gets shaded. Pick a convenient point on either side and see if that point's coordinates fit the given inequality. The point $(0, 0)$ is an easy one to work with. Do those coordinates satisfy the inequality?

$$y \le x + 2$$

$$0 \overset{?}{\le} 0 + 2$$

$$0 \overset{?}{\le} 2 \quad \text{Yes.}$$

So the point $(0, 0)$ must be on the shady side of the bent line, and the answer is (E).

Pick a Point.

When you're trying to find an equation that fits a graph, pick a point or two from the graph and try them in the equations.

Circles and Parabolas

The only curved graphs you'll find on a Level IC test are circles and parabolas. Like slope-intercept questions, these questions are essentially algebraic. Circles questions, like Example 5, are often just a matter of recalling and applying the appropriate equation.

Example 5

Which of the following equations describes the set of all points (x, y) in the coordinate plane that are a distance of 5 from the point $(-3, 4)$?

(A) $(x + 3) + (y - 4) = 5$

(B) $(x - 3) + (y + 4) = 5$

(C) $(x + 3)^2 + (y - 4)^2 = 5$

(D) $(x + 3)^2 + (y - 4)^2 = 25$

(E) $(x - 3)^2 + (y + 4)^2 = 25$

To use the formula for the equation of a circle, you need the coordinates (h, k) of the center. Here they're given: $(-3, 4)$. And you need the radius r. Here $r = 5$. So the equation is:

$$(x - h)^2 + (y - k)^2 = r^2$$
$$(x + 3)^2 + (y - 4)^2 = 25$$

The answer is (D).

There are all kinds of parabolas, and there's no simple, general parabola formula for you to memorize. You should know, however, that the graph of the general quadratic equation $y = ax^2 + bx + c$ is a parabola. It's one that opens up either on top or on bottom, with an axis of symmetry parallel to the y-axis. Here, for example, are parabolas that represent the equations $y = x^2 - 2x + 1$ and $y = -x^2 + 4$.

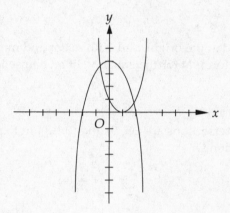

Circle

The equation of a circle centered at (h, k) and with radius r is:

$$(x - h)^2 + (y - k)^2 = r^2$$

Parabola

The graph of the general quadratic equation $y = ax^2 + bx + c$ is a parabola with an axis of symmetry parallel to the y-axis.

★Level IIC Topic

The following topic is tested only on the SAT II: Mathematics Level IIC Test. If you're preparing for Level IC, you won't need the information that follows. You're finished with your review of coordinate geometry topics. To see how much you've learned, try the questions in the Follow-Up Test at the end of the chapter. If you're preparing for the SAT II: Mathematics Level IIC Test, you'll need to know about the next topic, so read on!

 Ellipses and Hyperbolas

You might well encounter an ellipse on the Level IIC test. Example 6 tests your memory of the relevant formula.

Find Your Next Move

This topic relates to SAT II: Mathematics Level IIC only. If you're preparing for Level IC, skip this topic and move on to the Follow-Up Test at the end of the chapter.

*Example 6

Which of the following is an equation of an ellipse centered at the origin and with axial intersections at (0, ±3) and (±2, 0) ?

(A) $\dfrac{x}{2} + \dfrac{y}{3} = 1$

(B) $\dfrac{x}{2} + \dfrac{y}{3} = 2$

(C) $\dfrac{x}{3} + \dfrac{y}{2} = 2$

(D) $\dfrac{x^2}{4} + \dfrac{y^2}{9} = 1$

(E) $\dfrac{x^2}{4} + \dfrac{y^2}{9} = 2$

When centered at the origin, and with major and minor axes along the x- and y-axes, the formula for the equation of an ellipse is:

$$\dfrac{x^2}{a^2} + \dfrac{y^2}{b^2} = 1$$

where the axial intersections are (±a, 0) and (0, ±b). In Example 6, a = 2 and b = 3, so the equation is:

$$\dfrac{x^2}{2^2} + \dfrac{y^2}{3^2} = 1$$

$$\dfrac{x^2}{4} + \dfrac{y^2}{9} = 1$$

The answer is (D).

Note that if you don't remember the formula, you can still find the answer. It just takes a little longer. You have four points whose coordinates will satisfy the correct equation: (0, 3), (0, –3), (2, 0), and (–2, 0). The only choice that works with all of these points is (D).

Every once in a while, a hyperbola turns up on the Level IIC test. Here's an equation of a hyperbola that looks a lot like the equation for an ellipse, except that there's a minus sign.

$$\dfrac{x^2}{a^2} - \dfrac{y^2}{b^2} = 1$$

Now that you have seen how to solve these typical SAT II coordinate-geometry questions, it's time to try some more on your own. Take the following Coordinate Geometry Follow-Up Test.

Ellipse

The equation of an ellipse centered at the origin and with axial intersections at (±a, 0) and (0, ±b) is:

$$\dfrac{x^2}{a^2} + \dfrac{y^2}{b^2} = 1$$

Hyperbola

The equation of a hyperbola centered at the origin and with foci on the x-axis is:

$$\dfrac{x^2}{a^2} - \dfrac{y^2}{b^2} = 1$$

You've covered the coordinate geometry topics that you're likely to encounter on the SAT II: Mathematics Test. You've reviewed the facts and formulas, and learned some useful strategies. Now it's time for you to try another set of typical SAT II coordinate geometry questions.

Coordinate Geometry Follow-Up Test

Level IC: 5 Questions (6 Minutes)
Level IIC: 6 Questions (8 Minutes)

Directions for Level IC: If you are preparing for the SAT II: Mathematics Level IC Test, solve problems 1–5 and choose the best answer from those given. Fill in the oval corresponding to the best answer choice in the grid to the right of each question. (Answers and explanations begin on page 140.)

Directions for Level IIC: If you are preparing for the SAT II: Mathematics Level IIC Test, solve problems 1–6 and choose the best answer from those given. Fill in the oval corresponding to the best answer choice in the grid to the right of each question. (Answers and explanations begin on page 140.)

DO YOUR FIGURING HERE.

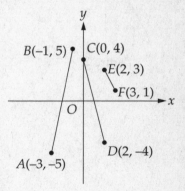

Figure 1

1. In Figure 1, if the midpoints of segments AB, CD, and EF are connected, what is the area of the resulting triangle?

    (A) 1

    (B) 1.5

    (C) 2

    (D) 2.5

    (E) 3   Ⓐ Ⓑ Ⓒ Ⓓ Ⓔ

2. If the line $y = 8x - 24$ intersects the line $y = mx + 12$ in the fourth quadrant, which of the following statements must be true?

    (A) $m < -4$

    (B) $-4 < m < 3$

    (C) $-3 < m < 3$

    (D) $0 < m < 4$

    (E) $m > 3$ Ⓐ Ⓑ Ⓒ Ⓓ Ⓔ

DO YOUR FIGURING HERE.

3. Which of the following lines is perpendicular to the line $y = -2x + 3$ and has the same y-intercept as the line $y = 2x - 3$?

    (A) $y = \dfrac{1}{2}x + 3$

    (B) $y = \dfrac{1}{2}x - 3$

    (C) $y = -\dfrac{1}{2}x + 3$

    (D) $y = 2x + 3$

    (E) $y = 2x - 3$ 

4. The shaded portion of Figure 2 shows the graph of which of the following?

    (A) $y(y - 2x) \geq 0$

    (B) $y(y - 2x) \leq 0$

    (C) $y\left(y - \dfrac{1}{2}x\right) \geq 0$

    (D) $y\left(y - \dfrac{1}{2}x\right) \leq 0$

    (E) $y\left(y + \dfrac{1}{2}x\right) \leq 0$ Ⓐ Ⓑ Ⓒ Ⓓ Ⓔ

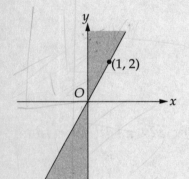

Figure 2

DO YOUR FIGURING HERE.

5.   The graph of the equation $x^2 + y^2 = 25$ includes how many points (x, y) in the coordinate plane where x and y are both integers?

(A)   Four

(B)   Five

(C)   Eight

(D)   Ten

(E)   Twelve

Ⓐ Ⓑ Ⓒ Ⓓ Ⓔ

★6.   Which of the following is a point at which the ellipse $\dfrac{x^2}{9} + \dfrac{y^2}{16} = 1$ intersects the y-axis?

(A) $(0, -3)$

(B) $(0, -4)$

(C) $(0, -8)$

(D) $(0, -9)$

(E) $(0, -16)$

Ⓐ Ⓑ Ⓒ Ⓓ Ⓔ

STOP!   END OF TEST. DO NOT TURN THE PAGE UNTIL YOU ARE READY TO CHECK YOUR ANSWERS.

Follow-Up Test—Answers and Explanations

Answer Key   1. *E*   2. *A*   3. *B*   4. *A*   5. *E*   6. *B*

1. **(E)**—First find the three midpoints.

midpoint of $AB = \left(\dfrac{-3+(-1)}{2},\ \dfrac{-5+5}{2}\right) = (-2,\ 0)$

midpoint of $CD = \left(\dfrac{0+2}{2},\ \dfrac{4+(-4)}{2}\right) = (1,\ 0)$

midpoint of $EF = \left(\dfrac{2+3}{2},\ \dfrac{3+1}{2}\right) = (2.5,\ 2)$

So the triangle looks like this.

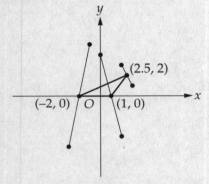

The base is 3 and the height is 2, so the area is $\frac{1}{2}(3)(2) = 3$.

2. **(A)**—The y-intercept of the line $y = 8x - 24$ is –24. Find the x-intercept by plugging in $y = 0$:

$$0 = 8x - 24$$
$$8x = 24$$
$$x = 3$$

The y-intercept of the line $y = mx + 12$ is +12. To intersect in the fourth quadrant—which is the lower right quadrant—this second line has to have a negative slope. Not just *any* negative slope, but one negative enough to get it down across the x-axis before it hits the other line. In other words, the x-intercept of this second line has to be somewhere

between the origin and the point (3, 0).

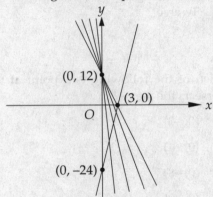

A line that goes from (0, 12) to (3, 0) would have a slope of $\dfrac{0-12}{3-0} = -4$, so the slope of the line $y = mx + 12$ must be *less* than –4.

3. **(B)**—A line that's perpendicular to $y = -2x + 3$ has a slope that's the negative reciprocal of –2, which is $\dfrac{1}{2}$. That narrows the choices to (A) and (B). The y-intercept of $y = -2x - 3$ is –3, and that's the y-intercept in (B).

4. **(A)**—Each of the answer choices is in the form of the product of two factors on the left and a "≥ 0" or "≤ 0" on the right. The product will be negative when the two factors have opposite signs, and it will be positive when the factors have the same sign. Choice (A), for example, has a "≥ 0," so you'll be looking for the factors to have the same sign.

Either:
$$y \geq 0 \quad \text{and} \quad y - 2x \geq 0$$
$$y \geq 0 \quad \text{and} \quad y \geq 2x$$

Or:
$$y \leq 0 \quad \text{and} \quad y - 2x \leq 0$$
$$y \leq 0 \quad \text{and} \quad y \leq 2x$$

The graph of $y \geq 0$ and $y \geq 2x$ looks like this:

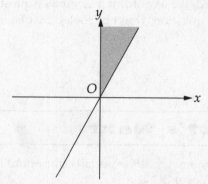

The graph of $y \leq 0$ and $y \leq 2x$ looks like this:

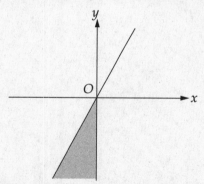

Together they make the graph in the figure.

5. **(E)**—The graph of $x^2 + y^2 = 25$ is a circle centered at the origin and with a radius of 5. The square of that radius can be the sum of the squares of integers in these cases on the axes.

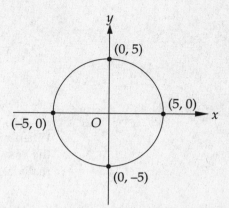

And x and y are both integers at these points as well.

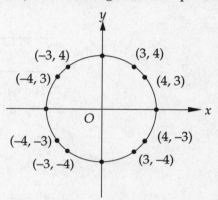

That's a total of 12. Remember, keep your eyes peeled for 3–4–5 triangles!

6. **(B)**—You might think at first glance that what you're supposed to do is graph the ellipse and somehow figure out where the ellipse crosses the y-axis. But hold on a second! What does a point of intersection represents in coordinate geometry? It represents a pair of numbers that satisfies both equations. The y-axis is the graph of the equation $x = 0$, so what you're looking for is a solution for both the equation $\dfrac{x^2}{9} + \dfrac{y^2}{16} = 1$ and the equation $x = 0$. Plug $x = 0$ into the ellipse equation and you get:

$$\frac{0^2}{9} + \frac{y^2}{16} = 1$$

$$\frac{y^2}{16} = 1$$

$$y^2 = 16$$

$$y = \pm 4$$

When x is 0, y is either 4 or –4, so the y-intercepts are (0, 4) and (0, –4), and the answer is (B). So, what looked like a coordinate geometry question was really a simultaneous equations question. Don't be fooled by disguises!

What's Next

The next chapter discusses trigonometry; it's especially important if you're taking the SAT II: Mathematics Level IIC Test.

CHAPTER 8

Trigonometry

Trigonometry is a relatively small topic on SAT II: Mathematics Level IC, but it's a huge topic on Level IIC. A typical Level IC test includes only about four trigonometry questions, which are generally among the more difficult questions in the latter half of the test. A typical Level IIC test, on the other hand, includes about ten trigonometry questions—that's one-fifth of the test. Bottom line: You could ignore the trigonometry questions and still get a good score on Level IC, but you have to be able to do at least *some* of the trigonometry questions to get a good score on Level IIC.

How to Use This Chapter

Maybe you already know all the trigonometry you need. You can find out by taking the Trigonometry Diagnostic Test on page 145. The first two questions on the Diagnostic Test are typical of what you could expect on either a Level IC or a Level IIC Mathematics Subject Test. Questions 3–6, though, could appear only on a Level IIC test. We've marked those Level IIC questions with stars. Check your answers using the answer key following the test. No matter how you score, don't worry! The answer key also shows where to find a detailed explanation for each question. The "Find Your Study Plan" section that follows the test will suggest next steps based on your performance on the Diagnostic.

Find Your Level

How you use this chapter really depends on which test you're taking and how much time you have to prep. Find your level and pace below.

Taking the Mathematics Level IC Test? Try the first two questions in the Trigonometry Diagnostic Test. Skip the Level IIC questions that we've marked with a star. Read the rest of the chapter, skipping any material that we've flagged as relating to Level IIC only. Then try the first two questions in the Follow-Up Test at the end of the chapter.

Trigonometry Facts and Formulas in This Chapter

- SOHCAHTOA (p. 149)
- Level IC Identities (p. 150)

Topics Relating to Level IIC only:

- Cotangent, Secant, and Cosecant (p. 151)★
- Level IIC Identities (p. 152)★
- Radians (p. 153)★
- Amplitude and Period (p. 155)★
- Law of Sines and Law of Cosines (p. 157)★

★
What's This Mean?

This icon appears next to more difficult topics and questions that would appear only on the Level IIC Test.

Kaplan Strategies in This Chapter

- IC Strategy: Set your calculator to degree mode and keep it there (p. 150).
- Keep an eye out for "$\sin^2 + \cos^2$" (p. 150).

Strategies Relating to Level IIC only:

- Make a mental picture of the graphs of $y = \sin x$, $y = \cos x$, and $y = \tan x$ (p. 154).
- Pick a point or two (p. 155).
- Don't be fooled by disguises (p. 156).

Mathematics Level IC Shortcut Try the first two questions in the Trigonometry Diagnostic Test. If you can answer both questions correctly, then you should probably skip ahead to the next chapter. If you're not too pressed for time, you could also try the first two questions in the Follow-Up Test at the end of the chapter.

Taking the Mathematics Level IIC Test? Do everything in this chapter. It's all relevant to the Level IIC Test.

Mathematics Level IIC Shortcut Take the Trigonometry Diagnostic Test and check your answers. The "Find Your Study Plan" section that follows the test will suggest next steps based on your Diagnostic Test score.

Panic Plan? If you're taking Level IC, skip this chapter. If you're taking Level IIC, make sure you can do at least some of the types of trigonometry questions that you see in this chapter.

Trigonometry Diagnostic Test

Level IC: 2 Questions (3 Minutes)
Level IIC: 6 Questions (8 Minutes)

Directions for Level IC: If you are preparing for the SAT II: Mathematics Level IC Test, solve problems 1–2 and choose the best answer from those given. Fill in the oval corresponding to the best answer choice in the grid to the right of each question. (Answers are on page 148.)

Directions for Level IIC: If you are preparing for the SAT II: Mathematics Level IIC Test, solve problems 1–6 and choose the best answer from those given. Fill in the oval corresponding to the best answer choice in the grid to the right of each question. (Answers are on page 148.)

DO YOUR FIGURING HERE.

1. In the right triangle in Figure 1, if $\theta = 39°$, what is the value of x ?

    (A)   9.9

    (B)   10.3

    (C)   11.3

    (D)   12.7

    (E)   13.9

    ⒶⒷⒸⒹⒺ

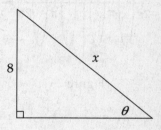

Figure 1

2. $(2\sin x)(3\sin x) + (6\cos x)(\cos x) =$

    (A)   1

    (B)   6

    (C)   12

    (D) $5\sin x + 7\cos x$

    (E) $6\sin x + 6\cos x$

    ⒶⒷⒸⒹⒺ

DO YOUR FIGURING HERE.

★3. If $\cos 2A = \frac{7}{19}$, what is the value of $\frac{1}{\cos^2 A - \sin^2 A}$?

(A) 0.18

(B) 0.37

(C) 0.74

(D) 1.36

(E) 2.71

Ⓐ Ⓑ Ⓒ Ⓓ Ⓔ

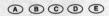

★4. Which of the following is an equation of the graph shown in Figure 2?

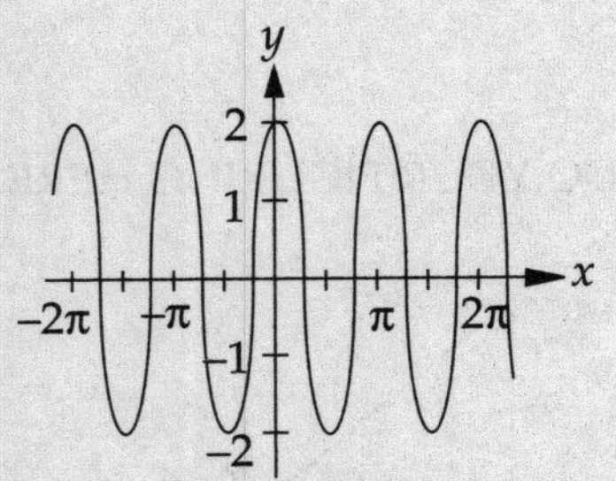

Figure 2

(A) $y = \frac{1}{2}\sin 2x$

(B) $y = \frac{1}{2}\cos 2x$

(C) $y = 2\sin\left(\frac{x}{2}\right)$

(D) $y = 2\cos\left(\frac{x}{2}\right)$

(E) $y = 2\cos 2x$

Ⓐ Ⓑ Ⓒ Ⓓ Ⓔ

★5. If $6\sin^2\theta - \sin\theta = 1$ and $0 \le \theta \le \pi$, what is the value of $\sin\theta$?

(A) $\frac{1}{6}$

(B) $\frac{1}{3}$

(C) $\frac{1}{2}$

(D) 19

(E) 30

Ⓐ Ⓑ Ⓒ Ⓓ Ⓔ

★6. What is the length of side *BC* in Figure 3?

DO YOUR FIGURING HERE.

(A)   7.3

(B)   7.7

(C)   8.1

(D)   8.5

(E)   8.9

Ⓐ Ⓑ Ⓒ Ⓓ Ⓔ

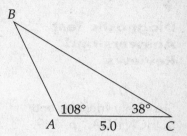

Figure 3

STOP!   END OF TEST. DO NOT TURN THE PAGE UNTIL
YOU ARE READY TO CHECK YOUR ANSWERS.

Diagnostic Test Answers and Reviews

1.   D
See "Right Triangles and SOHCAHTOA," p. 149.

2.   B
See "Level IC Identities," p. 150.

3.   E
See "Level IIC Identities," p. 151.*

4.   E
See "Graphing Trigonometric Functions," p. 153.*

5.   C
See "Solving Trigonometric Equations," p. 155.*

6.   D
See "Solving Triangles," p. 156.*

 Find Your Study Plan

The answer key shows where in this chapter to find explanations for the questions you missed. Here's how you should proceed based on your Diagnostic Test score.

Taking the Mathematics Level IC Test?

2: Superb! You're already good enough with trigonometry for the Level IC test. If you're on the Shortcut Plan, you might consider skipping this chapter. Or, if you want, you could just go straight to the Follow-Up Test at the end of the chapter.

1: Good. You're probably already good enough with trigonometry for the Level IC test. If you're on the Shortcut Plan, you might consider skipping this chapter. But you should at least look at the part of this chapter that discusses the question you did not get right. Then, if you want, you could just go straight to the Follow-Up Test at the end of the chapter.

0: Trigonometry may be a problem area for you, so you'd better spend some time with this chapter.

Taking the Mathematics Level IIC Test?

6: Superb! You're already good enough with trigonometry for the Level IIC test. If you're taking a "shortcut," you might consider skipping this chapter. Or you could just go straight to the Follow-Up Test at the end of the chapter.

4–5: Good. You're probably already good enough with trigonometry for the Level IIC test. If you're taking a "shortcut," you might consider skipping this chapter. But you should at least look at the parts of this chapter that discuss any questions you did not get right. Then, if you have time, you could move on to the Follow-Up Test at the end of the chapter.

0–3: Trigonometry may be a problem area for you, so you'd better spend some time with this chapter.

Trigonometry Test Topics

As the Trigonometry Diagnostic Test showed you, there's a big difference between Level IC trigonometry and Level IIC trigonometry. If you're taking Level IC, there's not a whole lot you have to remember. All you really need to know about is the sine, cosine, and tangent of acute angles measured in degrees, and a few basic trigonometric identities. We'll use the first two questions from the Diagnostic Test to review these basic topics.

If you're taking Level IIC, however, you have a lot of definitions and identities to remember. In addition to the Level IC topics above, you also need to know about: cotangent, secant, and cosecant; angles greater than 90° and angles measured in radians; graphing trigonometric functions; arcsin, arccos, and arctan; half-angle and double-angle identities; and the laws of sines and cosines. We'll use the last four questions on the diagnostic to review these more advanced trigonometry topics.

 ## Right Triangles and SOHCAHTOA

To remember the definitions of sine, cosine, and tangent as they apply to right triangles, use the mnemonic SOHCAHTOA. That and a calculator are all you need to answer Example 1:

Example 1

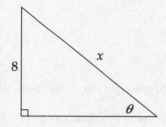

Figure 1

In the right triangle in Figure 1, if $\theta = 39°$, what is the value of x ?

(A)   9.9

(B)   10.3

(C)   11.3

(D)   12.7

(E)   13.9

The 39° angle is opposite the given 8, and the side you're looking for is the hypotenuse, so you can use the sine to find x.

$$\text{sine} = \frac{\text{opposite}}{\text{hypotenuse}}$$
$$\sin 39° = \frac{8}{x}$$
$$x = \frac{8}{\sin 39°}$$

Test Topics

This icon appears next to each discussion of a math topic that's tested on the SAT II.

SOHCAHTOA

$$\text{Sine} = \frac{\text{Opposite}}{\text{Hypotenuse}}$$

$$\text{Cosine} = \frac{\text{Adjacent}}{\text{Hypotenuse}}$$

$$\text{Tangent} = \frac{\text{Opposite}}{\text{Adjacent}}$$

Level IC Strategy: Set Your Calculator to Degree Mode and Keep It There.

All angle measures on the Level IC test are in degrees, so make sure your calculator is in degree mode. Most angle measures on the Level IIC test are in degrees, but sometimes they're in radians, so read the questions carefully and switch your calculator to radian mode when appropriate.

Keep an Eye Out for "sin² + cos²."

For any angle x, take the sine and cosine, square them both, and the squares will add up to 1. This identity turns up a lot on both levels of the test.

Level IC Identities

$$\tan x = \frac{\sin x}{\cos x}$$

$$\sin^2 x + \cos^2 x = 1$$

Now, nobody expects you to know the sine of 39° off the top of your head. This is one of those places where you *have* to use your calculator. Punch in "8 ÷ sin 39 =," making sure your calculator is set to degree mode, and you'll get something like: 12.712125825, which is close to (D) 12.7.

 ### Level IC Identities

On the Level IC test you'll probably come across a basic trigonometric identities question like Example 2.

Example 2

$(2\sin x)(3\sin x) + (6\cos x)(\cos x) =$

(A) 1

(B) 6

(C) 12

(D) $5\sin x + 7\cos x$

(E) $6\sin x + 6\cos x$

Start by multiplying to get rid of the parentheses, and see where that will lead you:
$$(2\sin x)(3\sin x) + (6\cos x)(\cos x) = 6\sin^2 x + 6\cos^2 x$$

When you see a sin², a plus sign, and a cos², a little bell should ring in your head. Remember this key trigonometric identity:

$$\sin^2 x + \cos^2 x = 1$$

For any angle x, take the sine and cosine, square them both, and the squares will add up to 1. (This relationship is really just a variation of the Pythagorean theorem.) The test makers love this identity—it turns up a lot on both Levels IC and IIC. So keep your eye out for it.

Here in Example 2, when you spot "6sin² x + 6cos² x," you should think immediately about how to extract "sin² x + cos² x" from it. You do that by factoring out a 6.

$$6\sin^2 x + 6\cos^2 x = 6\left(\sin^2 x + \cos^2 x\right)$$
$$= 6(1)$$
$$= 6$$

The answer is (B).

★Level IIC Topics

The rest of the topics in this chapter are tested only on the SAT II: Mathematics Level IIC Test. If you're preparing for Level IC, you're finished with your review of trigonometry topics. To see how much you've learned, try the first two questions in the Follow-Up Test at the end of the chapter.

If you're preparing for the SAT II: Mathematics Level IIC Test, you'll need to read on. For Level IIC you have to know a lot more than just SOHCAHTOA and "$\sin^2 x + \cos^2 x = 1$." To begin with, you're supposed to know the other three trigonometric functions: cotangent, secant, and cosecant. Like the three basic functions (sine, cosine, and tangent), these other three functions can be defined in terms of "opposite," "adjacent," and "hypotenuse." They can also be defined as reciprocals of the three basic functions.

 ## Level IIC Identities

You're expected to know a lot more identities if you're taking the Level IIC test. Besides questions that ask explicitly about such identities, there are other questions that become a lot easier and faster to answer when you know the right identities. That's the case with Example 3:

★**Example 3**

If $\cos 2A = \dfrac{7}{19}$, what is the value of $\dfrac{1}{\cos^2 A - \sin^2 A}$?

(A) 0.18

(B) 0.37

(C) 0.74

(D) 1.36

(E) 2.71

 ## Find Your Next Move

These topics relate to SAT II: Mathematics Level IIC only. If you're preparing for Level IC, skip these topics and move on to the Follow-Up Test at the end of the chapter.

 ## Cotangent, Secant, and Cosecant

Cotangent =

$$\frac{\text{Adjacent}}{\text{Opposite}} = \frac{1}{\text{Tangent}}$$

Secant =

$$\frac{\text{Hypotenuse}}{\text{Adjacent}} = \frac{1}{\text{Cosine}}$$

Cosecant =

$$\frac{\text{Hypotenuse}}{\text{Opposite}} = \frac{1}{\text{Sine}}$$

Level IIC Identities

Functions of Sums

$$\sin(A + B) = \sin A \cos B + \cos A \sin B$$

$$\cos(A + B) = \cos A \cos B - \sin A \sin B$$

$$\tan(A + B) = \frac{\tan A + \tan B}{1 - \tan A \tan B}$$

Double-Angle Identities

$$\sin 2x = 2\sin x \cos x$$

$$\cos 2x = \cos^2 x - \sin^2 x$$
$$= 1 - 2\sin^2 x$$
$$= 2\cos^2 x - 1$$

$$\tan 2x = \frac{2\tan x}{1 - \tan^2 x}$$

Half-Angle Identities

$$\sin \frac{1}{2} A = \pm \sqrt{\frac{1 - \cos A}{2}}$$

$$\cos \frac{1}{2} A = \pm \sqrt{\frac{1 + \cos A}{2}}$$

$$\tan \frac{1}{2} A = \pm \sqrt{\frac{1 - \cos A}{1 + \cos A}}$$

Here's one way to answer this question. First find A :

$$\cos 2A = \frac{7}{19}$$

$$2A = \arccos \left(\frac{7}{19} \right)$$
$$2A \approx 68.38°$$
$$A \approx 34.19°$$

Then plug $A = 34.19$ into the expression you want to evaluate:

$$\frac{1}{\cos^2 A - \sin^2 A} = \frac{1}{(\cos 34.19°)^2 - (\sin 34.19°)^2}$$

$$\approx \frac{1}{(0.827)^2 - (0.562)^2}$$

$$\approx \frac{1}{0.684 - .316}$$

$$= \frac{1}{.368}$$

$$\approx 2.72$$

The answer is (E), but the above method is much longer, more involved, and more open to calculator error than necessary. The best way to answer this question is to recognize the relationship between what's given and what's asked for. Answering the question involves just one quick calculation if you remember the relevant identity:

$$\cos 2A = \cos^2 A - \sin^2 A$$

So the given $\cos 2A = \frac{7}{19}$ is the same as the denominator of the expression you're solving for:

$$\frac{1}{\cos^2 A - \sin^2 A} = \frac{1}{\cos 2A} = \frac{1}{\frac{7}{19}} = \frac{19}{7}$$

So the only calculating you need to do is: $19 \div 7 \approx 2.71$.

So it can come in handy to know lots of trigonometric identities. You don't absolutely *have* to know the double-angle and half-angle identities—there's usually a way around them—but you'll be better equipped to work quickly and efficiently if you do know them.

Graphing Trigonometric Functions

Radians

Example 4 is about graphing a trigonometric function:

π radians = 180 degrees

★Example 4

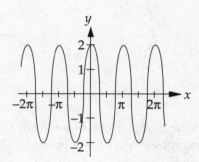

Figure 2

Which of the following is an equation of the graph shown in Figure 2?

(A) $\quad y = \dfrac{1}{2}\sin 2x$

(B) $\quad y = \dfrac{1}{2}\cos 2x$

(C) $\quad y = 2\sin\left(\dfrac{x}{2}\right)$

(D) $\quad y = 2\cos\left(\dfrac{x}{2}\right)$

(E) $\quad y = 2\cos 2x$

To graph a trigonometric function, put the angles (usually in radians) on the x-axis and the results of applying the function on the y-axis. You can always use your graphing calculator (if you have one and know how to use it) to find the graph of a trigonometric function. That would be one way to do Example 4.

But ultimately you're better off if you can have a true understanding, a "feeling," for what the graphs of trigonometric functions look like. The way to do that is to start with a picture in your mind of what the graphs of the three basic functions look like.

Level IIC Strategy: Make a Mental Picture of the Graphs of y = sin x, y = cos x, and y = tan x.

Fix in your mind images of the sine, cosine, and tangent graphs and you'll find that graphing trigonometric functions in general is easy and makes sense.

Basic trig function 1: The graph of $y = \sin x$ is a curve that goes through the origin because $\sin 0 = 0$, and rises to a crest at $(\frac{\pi}{2}, 1)$ because $\sin \frac{\pi}{2} = 1$:

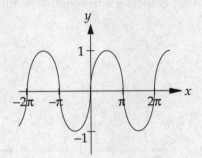

Basic trig function 2: The graph of $y = \cos x$ is a curve that looks just like the sine curve except it crosses the y-axis at $(0, 1)$ because $\cos 0 = 1$, falls to cross the x-axis at $(\frac{\pi}{2}, 0)$ because $\cos \frac{\pi}{2} = 0$, and continues on down to its floor at $(\pi, -1)$ because $\cos \pi = -1$:

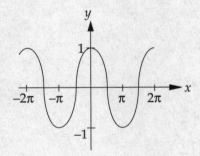

Basic trig function 3: The graph of $y = \tan x$ looks like this:

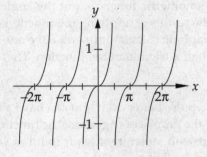

Many other graphs of trigonometric functions can be viewed as variations on these. The graph in Example 4, for instance, looks a lot like the cosine curve—it has a crest on the y-axis—but its maximum and minimum are ±2, and it falls to cross the x-axis at $\frac{\pi}{4}$, and bottoms out at $x = \frac{\pi}{2}$. In other words, it's the cosine curve with twice the amplitude and half the period. The amplitude is affected by the number in front of the "cos." Twice the amplitude means that number is 2. The period is affected by the number in front of the x. Half the period means the coefficient of x is 2. So the equation is $y = 2\cos 2x$. And the answer is (E).

Of course, you can always do a question like Example 4 by picking a point or two. The point (0, 2) satisfies equations (D) and (E) only. And of those choices, (E)'s the only one that works with the point $\left(\frac{\pi}{2}, -2\right)$.

 ## Solving Trigonometric Equations

Here's a question that looks like a complicated trigonometry question, but turns out not to be trigonometry at all:

★Example 5

If $6\sin^2\theta - \sin\theta = 1$ and $0 \le \theta \le \pi$, what is the value of $\sin\theta$?

(A) $\dfrac{1}{6}$

(B) $\dfrac{1}{3}$

(C) $\dfrac{1}{2}$

(D) 19

(E) 30

This sure looks like a trigonometry question. There are sines and thetas all over the place. But in fact this is just a plain old algebra question in disguise. You never need to find the sine or the arcsine of anything. You don't need to find θ. This is really just a quadratic equation where the unknown is written "sin θ" instead of the more usual "x." Perhaps you'll find the equation easier to deal with if you first replace "sin θ" with x :

$$6\sin^2\theta - \sin\theta = 1$$
$$6x^2 - x = 1$$

Amplitude and Period

For functions in the form

$y = a \sin bx$ or $y = a \cos bx$,

the **amplitude** of the curve is a and the **period** of the curve is $\dfrac{360}{b}$ degrees or $\dfrac{2\pi}{b}$ radians.

Level IIC Strategy: Pick a Point or Two.

To identify the equation for a particular graph, pick a point or two from the graph and try the coordinates in the equations.

Don't Be Fooled by Disguises.

Sometimes what looks like a trigonometry question turns out to be primarily an algebra question.

Now solve for x as usual:

$$6x^2 - x = 1$$
$$6x^2 - x - 1 = 0$$
$$(2x-1)(3x+1) = 0$$
$$2x-1=0 \text{ or } 3x+1=0$$
$$x = \frac{1}{2} \text{ or } -\frac{1}{3}$$

It's given that $0 \le \theta \le \pi$, so the sine is nonnegative and the answer is (C) $\frac{1}{2}$.

 ## Solving Triangles

You just saw an example that was an algebra question disguised as a trigonometry question. Example 6 is a trigonometry question disguised as a geometry question.

★Example 6

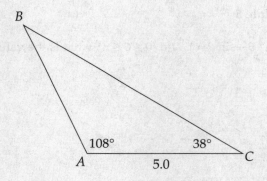

Figure 3

What is the length of side BC in Figure 3 ?

    (A)   7.3

    (B)   7.7

    (C)   8.1

    (D)   8.5

    (E)   8.9

You might not think that this is a trigonometry question at first. There's no "sin," "cos," or any other explicit trig function mentioned in the stem. And the triangle's not a right triangle. But trig is the tool you have to use to answer this question. This is a case of "solving a triangle," that is, finding the length of one or more sides.

With the Law of Sines and the Law of Cosines and a calculator, you can solve almost any triangle. If you know any two angles—which of course means that you know all three—and one side, you can use the Law of Sines to find the other two sides. If you know two sides and the angle between them—remember, it must be the angle *between* them—you can use the Law of Cosines to find the third side.

In Example 6, what you're given is two angles and a side, so you'll use the Law of Sines, which says simply that the sines are proportional to the opposite sides. Here the side you're looking for, *BC*, is opposite the 108°. The side you're given, *AC*, is opposite the unlabeled angle, which measures $180 - 108 - 38 = 34$ degrees. Now you can set up the proportion.

$$\frac{BC}{\sin A} = \frac{AC}{\sin B}$$

$$\frac{BC}{\sin 108°} = \frac{5.0}{\sin 34°}$$

$$BC = \frac{5\sin 108°}{\sin 34°} \approx 8.5$$

The answer is (D).

Now that you've reviewed all the relevant Trigonometry facts and formulas, seen some of the test makers' favorite trig situations, and learned a few essential Kaplan strategies, it's time to take another crack at some test-like trig questions. Try the Trigonometry Follow-Up Test.

Law of Sines and Law of Cosines

If you know any two angles and one side of a triangle, you can figure out the other sides by using the **Law of Sines**. For any triangle, the side lengths are proportional to the sines of the opposite angles:

$$\frac{a}{\sin A} = \frac{b}{\sin B} = \frac{c}{\sin C}$$

If you know two sides of a triangle and the angle between them, you can figure out the third side by using the **Law of Cosines**, which is a more general version of the Pythagorean theorem:

$$c^2 = a^2 + b^2 - 2ab\cos C$$

Trigonometry Follow-Up Test

Level IC: 2 Questions (3 Minutes)
Level IIC: 6 Questions (8 Minutes)

Directions for Level IC: If you are preparing for the SAT II: Mathematics Level IC Test, solve problems 1–2 and choose the best answer from those given. Fill in the oval corresponding to the best answer choice in the grid to the right of each question. (Answers and explanations begin on page 160.)

Directions for Level IIC: If you are preparing for the SAT II: Mathematics Level IIC Test, solve problems 1–6 and choose the best answer from those given. Fill in the oval corresponding to the best answer choice in the grid to the right of each question. (Answers and explanations begin on page 160.)

DO YOUR FIGURING HERE.

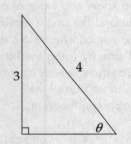

Figure 1

1. In Figure 1, what is the value of cos θ ?

    (A)   0.25

    (B)   0.66

    (C)   0.75

    (D)   0.80

    (E)   1.25

2. Where defined, $\dfrac{\cos^4 \theta - \sin^4 \theta}{\cos^2 \theta - \sin^2 \theta} =$

    (A)   1

    (B)   2

    (C)   cos θ + sin θ

    (D)   cos θ – sin θ

    (E)   cos 2θ + sin 2θ

DO YOUR FIGURING HERE.

★3. If $\sin\theta = \dfrac{1}{2}\cos\theta$, and $0 \le \theta \le \dfrac{\pi}{2}$, what is the value of $\dfrac{1}{2}\sin\theta$?

(A) 0.22

(B) 0.25

(C) 0.45

(D) 0.50

(E) 0.75

Ⓐ Ⓑ Ⓒ Ⓓ Ⓔ

★4. If θ is an acute angle for which $\tan^2\theta = 6\tan\theta - 9$, what is the degree measure of θ ?

(A) 51.3

(B) 60.0

(C) 71.6

(D) 79.7

(E) 83.5

Ⓐ Ⓑ Ⓒ Ⓓ Ⓔ

★5. What is the y-coordinate of the point at which the graph of $y = 2\sin x - \cos 2x$ intersects the y-axis?

(A) –2   (B) –1   (C) 0   (D) 1   (E) 2

Ⓐ Ⓑ Ⓒ Ⓓ Ⓔ

★6. Figure 2 shows a regular pentagon with all of its vertices on the sides of a rectangle. If $BC = \dfrac{1}{4}$, what is the perimeter of the pentagon?

(A) 0.31

(B) 0.62

(C) 0.77

(D) 0.80

(E) 1.00

Ⓐ Ⓑ Ⓒ Ⓓ Ⓔ

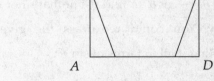

Figure 2

STOP!   **END OF TEST. DO NOT TURN THE PAGE UNTIL YOU ARE READY TO CHECK YOUR ANSWERS.**

Follow-Up Test—Answers and Explanations

Answer Key   1. *B*   2. *A*   3. *A*   4. *C*   5. *B*   6. *C*

1. **(B)**—Cosine is adjacent over hypotenuse. You'll have to use the Pythagorean theorem to find the adjacent leg:

$$\text{leg} = \sqrt{4^2 - 3^2}$$
$$= \sqrt{16 - 9}$$
$$= \sqrt{7}$$

$$\cos\theta = \frac{\text{adjacent}}{\text{hypotenuse}} = \frac{\sqrt{7}}{4} \approx 0.66$$

2. **(A)**—The numerator is the difference of squares and so can be factored:

$$\frac{\cos^4\theta - \sin^4\theta}{\cos^2\theta - \sin^2\theta}$$

$$= \frac{\left(\cos^2\theta + \sin^2\theta\right)\left(\cos^2\theta - \sin^2\theta\right)}{\cos^2\theta - \sin^2\theta}$$

$$= \cos^2\theta + \sin^2\theta$$

The "$\cos^2\theta + \sin^2\theta$" should look familiar—it equals 1.

★3. **(A)**—You could use the given equation $\sin\theta = \frac{1}{2}\cos\theta$ to find θ, but that's not really necessary. You could reexpress the given equation putting it all in terms of $\sin\theta$:

$$\sin\theta = \frac{1}{2}\cos\theta$$

$$\sin\theta = \frac{1}{2}\sqrt{1 - \sin^2\theta}$$

$$(\sin\theta)^2 = \left(\frac{1}{2}\sqrt{1 - \sin^2\theta}\right)^2$$

$$\sin^2\theta = \frac{1}{4}\left(1 - \sin^2\theta\right)$$

$$4\sin^2\theta = 1 - \sin^2\theta$$

$$5\sin^2\theta = 1$$

$$\sin^2\theta = \frac{1}{5}$$

$$\sin\theta = \sqrt{\frac{1}{5}} \approx 0.45$$

$$\frac{1}{2}\sin\theta \approx 0.22$$

★4. **(C)**—This is just a quadratic equation where the unknown is $\tan\theta$:

$$\tan^2\theta = 6\tan\theta - 9$$

$$\tan^2\theta - 6\tan\theta + 9 = 0$$

$$(\tan\theta - 3)^2 = 0$$

$$\tan\theta = 3$$

$$\theta = \arctan(3)$$

$$\approx 71.6°$$

★5. **(B)**—The graph intersects the y-axis at the point where $x = 0$:

$$y = 2\sin x - \cos 2x$$

$$= 2\sin(0) - \cos(0)$$

$$= 2(0) - 1$$

$$= -1$$

★6. (C)—Mark up the figure. The angles of the rectangle are right angles. To figure out the interior angles of the pentagon, use the formula for the degree measure of each interior angle of a regular polygon of n sides:

$$\text{Interior angle} = \frac{(n-2)180}{n}$$
$$= \frac{(5-2)180}{5}$$
$$= 108$$

Label all the angles you can:

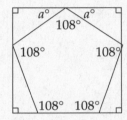

The angles marked $a°$ are equal and together with 108° add up to 180°:

$$108 + 2a = 180$$
$$2a = 72$$
$$a = 36$$

So the right triangle in the upper left looks like this:

The leg on top (opposite the 54° angle) is half of BC, or $\frac{1}{8}$. The hypotenuse of this right triangle is one side of the pentagon whose perimeter you are looking for, so call the hypotenuse x:

You have the leg that's *opposite* the 54° and you're looking for the *hypotenuse*, so use the *sine*:

$$\sin 54° = \frac{\frac{1}{8}}{x}$$
$$x = \frac{\frac{1}{8}}{\sin 54°} \approx 0.155$$

That's not the answer, however. That's just one side of the pentagon, and you want the perimeter, so multiply that side length by 5:

$$\text{Perimeter} \approx 5 \times 0.155 \approx 0.77$$

What's Next

The next chapter discusses functions, an important category of questions on both the Mathematics Level IC Test and the Mathematics Level IIC Test.

CHAPTER 9

Functions

Functions is a fairly large category on the SAT II: Mathematics Level IC test, and a very large category on the Level IIC test. The Level IC test includes about six functions questions, and the Level IIC test includes about twelve. Whichever test you're taking, but especially if you're taking Level IIC, you need to be comfortable with the material in this chapter.

How to Use This Chapter

Maybe you already know everything you need to know about functions. You can find out by taking the Functions Diagnostic Test on page 165. The first four questions on the Diagnostic Test are typical of what you could expect on either a Level IC or a Level IIC Mathematics Subject Test. Questions 5–6, though, would appear only on a Level IIC test. We've marked those Level IIC questions with stars. Check your answers using the answer key following the test. No matter how you score, don't worry! The answer key also shows where to find a detailed explanation for each question. The "Find Your Study Plan" section that follows the test will suggest next steps based on your performance on the Diagnostic.

 Find Your Level

How you use this chapter really depends on which test you're taking and how much time you have to prep. Find your level and pace below.

Taking the Mathematics Level IC Test? Try the first four questions in the Functions Diagnostic Test. Skip the Level IIC questions that we've marked with a star. Read the rest of the chapter, skipping any material that we've flagged as relating to Level IIC only. Then try the first four questions in the Follow-Up Test at the end of the chapter.

Mathematics Level IC Shortcut Try the first four questions in the Functions Diagnostic Test and check your answers. The "Find Your Study

 Functions Facts and Formulas in This Chapter

- Functions: Domain and Range (p. 169)
- Compound Functions (p. 170)
- Maximums and Minimums (p. 171)

Topics Relating to Level IIC only:
- Undefined (p. 174)★
- Inverse Functions (p. 175)★

★
What's This Mean?

This icon appears next to more difficult topics and questions that would appear only on the Level IIC Test.

Kaplan Strategies in This Chapter

- Don't be afraid of functions (p. 169).
- Watch those parentheses (p. 171).

Plan" section that follows the test will suggest next steps based on your Diagnostic Test score.

Taking the Mathematics Level IIC Test? Do everything in this chapter. It's all relevant to the Level IIC Test.

Mathematics Level IIC Shortcut Take the Functions Diagnostic Test and check your answers. The "Find Your Study Plan" section that follows the test will suggest next steps based on your Diagnostic Test score.

Panic Plan? Look through the chapter quickly and make sure you're comfortable with the material. If you're not comfortable with functions, especially if you're taking Level IIC, spend some time with this critical chapter.

Functions Diagnostic Test

Level IC: 4 Questions (6 Minutes)
Level IIC: 6 Questions (8 Minutes)

Directions for Level IC: If you are preparing for the SAT II: Mathematics Level IC Test, solve problems 1–4 and choose the best answer from those given. Fill in the oval corresponding to the best answer choice in the grid to the right of each question. (Answers are on page 168.)

Directions for Level IIC: If you are preparing for the SAT II: Mathematics Level IIC Test, solve problems 1–6 and choose the best answer from those given. Fill in the oval corresponding to the best answer choice in the grid to the right of each question. (Answers are on page 168.)

DO YOUR FIGURING HERE.

1. If $f(x) = x^2 + \dfrac{x}{2}$, then $f(a + 2) =$

    (A) $a^2 + \dfrac{a}{2}$

    (B) $a^2 + \dfrac{5a}{2} + 2$

    (C) $a^2 + \dfrac{5a}{2} + 5$

    (D) $a^2 + \dfrac{9a}{2} + 2$

    (E) $a^2 + \dfrac{9a}{2} + 5$

    Ⓐ Ⓑ Ⓒ Ⓓ Ⓔ

2. If $f(x) = \sqrt{x}$ and $g(x) = \sqrt{x^2 + 4}$, what is the value of $f(g(2))$?

    (A) 0

    (B) 1.41

    (C) 1.68

    (D) 2.45

    (E) 2.83

    Ⓐ Ⓑ Ⓒ Ⓓ Ⓔ

DO YOUR FIGURING HERE.

3. What is the maximum value of $f(x) = 2 - (x + 2)^2$?

(A) −4

(B) −2

(C)   0

(D)   2

(E)   4

Ⓐ Ⓑ Ⓒ Ⓓ Ⓔ

4. If $f(x) = |1 - x|$, which of the following could be the graph of $y = f(x)$?

(A)
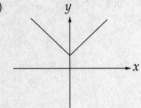

(B)
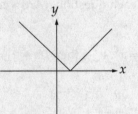

(C)
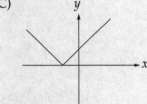

(D)
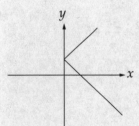

(E)
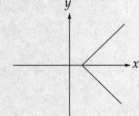

Ⓐ Ⓑ Ⓒ Ⓓ Ⓔ

DO YOUR FIGURING HERE.

★5. If $f(x) = \dfrac{\sqrt{x^2 - 4}}{x - 4}$, what are all the values of x for which $f(x)$ is defined?

(A) All real numbers except 4

(B) All real numbers except –2 and 2

(C) All real numbers greater than or equal to –2 and less than or equal to 2

(D) All real numbers less than or equal to –2 or greater than or equal to 2

(E) All real numbers less than or equal to –2 or greater than or equal to 2, except 4

Ⓐ Ⓑ Ⓒ Ⓓ Ⓔ

★6. If $f(x) = \dfrac{1}{3}x + 3$, then $f^{-1}(x) =$

(A) $-\dfrac{1}{3}x - 3$

(B) $-3x + \dfrac{1}{3}$

(C) $3x + \dfrac{1}{3}$

(D) $3x - 3$

(E) $3x - 9$

Ⓐ Ⓑ Ⓒ Ⓓ Ⓔ

STOP! **END OF TEST. DO NOT TURN THE PAGE UNTIL YOU ARE READY TO CHECK YOUR ANSWERS.**

Diagnostic Test Answers and Reviews

1.   E
See "Substitution," p. 169.

2.   C
See "Compound Functions," p. 170.

3.   D
See "Maximums and Minimums," p. 171.

4.   B
See "Graphing Functions," p. 172.

5.   E
See "Undefined," p. 173.★

6.   E
See "Inverse Functions," p. 174.★

Find Your Study Plan

The answer key shows where in this chapter to find explanations for the questions you missed. Here's how you should proceed based on your Diagnostic Test score.

Taking the Mathematics Level IC Test?

4: Superb! You're already good enough with functions for the Level IC test. If you're taking a "shortcut," you might consider skipping this chapter. Or, if you want, you could just go straight to the Follow-Up Test at the end of the chapter.

3: Good. You're probably already good enough with functions for the Level IC test. If you're taking a "shortcut," you might consider skipping this chapter. But you should at least look at the part of this chapter that discusses the question you did not get right. Then, if you have time, you could go straight to the Follow-Up Test at the end of the chapter.

0–2: Functions may be a problem area for you, so you'd better spend some time with this chapter.

Taking the Mathematics Level IIC Test?

6: Superb! You're already good enough with functions for the Level IIC test. If you're taking a "shortcut," you might consider skipping this chapter. Or you could just go straight to the Follow-Up Test at the end of the chapter.

4–5: Good. You're probably already good enough with functions for the Level IIC test. If you're taking a "shortcut," you might consider skipping this chapter. But you should at least look at the parts of this chapter that discuss any questions you did not get right. Then, if you have time, you could move on to the Follow-Up Test at the end of the chapter.

0–3: Functions may be a problem area for you, so you'd better spend some time with this chapter.

Functions Test Topics

We'll use the questions on the Functions Diagnostic Test to illustrate the conventions of functions. Only the first four Diagnostic questions are relevant to the Mathematics Level IC test. If you're preparing for the Level IC Test, you can skip the discussion of the last two questions. If you're preparing for Level IIC, however, each topic will be relevant to you.

Who's Afraid of Functions?

Lots of students are afraid of functions. But there's nothing especially difficult about them. They just *look* scary. Once you "get" the conventions, however, you'll never be afraid of functions again. Here's a quick and painless review of the basic things you need to know about functions.

A function is a process that turns a number into another number. Squaring is an example of a function. For any number you can think of, there is a unique number that is its square. The conventional way of writing this function is:

$$f(x) = x^2$$

When you apply the function to some particular number, such as –5, you write it this way:

$$f(-5)$$

And to find the value of $f(-5)$, you plug $x = -5$ into the definition:

$$f(x) = x^2$$
$$f(-5) = (-5)^2 = 25$$

 ## Substitution

The most straightforward functions questions you'll encounter on the SAT II are questions like Example 1 that ask you simply to apply a function to some number or expression.

Example 1

If $f(x) = x^2 + \dfrac{x}{2}$, then $f(a + 2) =$

(A) $a^2 + \dfrac{a}{2}$

(B) $a^2 + \dfrac{5a}{2} + 2$

(C) $a^2 + \dfrac{5a}{2} + 5$

(D) $a^2 + \dfrac{9a}{2} + 2$

(E) $a^2 + \dfrac{9a}{2} + 5$

To find $f(a + 2)$, plug $x = a + 2$ into the definition. Substituting an algebraic expression for x is a little more complicated than substituting a number for x, but the idea's the same.

 Test Topics

This icon appears next to each discussion of a math topic that's tested on the SAT II.

 Functions: Domain and Range

A **function** is a set of ordered pairs (x, y) such that for each value of x there is one and only one value of y.

The set of all allowable x values is called the **domain**. Note that, according to the SAT II directions, the domain of any function f is assumed to be the set of all real numbers x for which $f(x)$ is a real number.

The corresponding set of all y values is called the **range**.

 Don't Be Afraid of Functions.

There's nothing inherently difficult about functions. There are not a lot of formulas or theorems to remember. Getting comfortable with functions just means understanding the symbolism and conventions.

$$f(x) = x^2 + \frac{x}{2}$$

$$f(a+2) = (a+2)^2 + \frac{a+2}{2}$$

$$= a^2 + 4a + 4 + \frac{a}{2} + 1$$

$$= a^2 + \frac{9a}{2} + 5$$

The answer is (E).

 ## Compound Functions

The letter f is not the only letter used to designate a function, though it's the most popular. Second in popularity is the letter g, which is generally used in a question that includes two different functions. Take a look at Example 2:

Example 2

If $f(x) = \sqrt{x}$ and $g(x) = \sqrt{x^2 + 4}$, what is the value of $f(g(2))$?

(A)   0

(B)   1.41

(C)   1.68

(D)   2.45

(E)   2.83

When one function is written inside another function's parentheses, apply the inside function first:

$$g(x) = \sqrt{x^2 + 4}$$

$$g(2) = \sqrt{2^2 + 4} = \sqrt{8}$$

Then apply the outside function to the result:

$$f(x) = \sqrt{x}$$

$$f(\sqrt{8}) = \sqrt{\sqrt{8}} \approx 1.68$$

The answer is (C).

Notice that there's a difference between $f(g(2))$ and $g(f(2))$. In the latter, you're to apply the function f first, and then the function g to the result. The order makes a difference: $f(g(2)) \approx 1.68$, but $g(f(2)) \approx 2.45$ (which is of course included as a distractor).

Notice also that there's a difference between $f(g(x))$ and $f(x)g(x)$. In the former expression, a hierarchy is indicated: The function g is inside the parentheses of function f, so you apply g first, and then you apply f to the result. In the expression $f(x)g(x)$, however, the two functions are written side by side, which indicates multiplication. You apply the function f to 2, and you apply the function g to 2, and then you multiply the results:

$$f(2) = \sqrt{2}$$
$$g(2) = \sqrt{8}$$
$$f(2)g(2) = \left(\sqrt{2}\right)\left(\sqrt{8}\right) = \sqrt{16} = 4$$

So when it comes to functions, be sure to pay close attention to order and parentheses.

 ## Maximums and Minimums

Another typical functions question is one that asks for a minimum or, as in Example 3, maximum value of a function.

Example 3

What is the maximum value of $f(x) = 2 - (x + 2)^2$?

(A) −4

(B) −2

(C) 0

(D) 2

(E) 4

If you have a graphing calculator (and know how to use it), you could graph the function and trace the graph to find the maximum. But it's really a lot easier if you conceptualize the situation. The expression $2 - (x + 2)^2$ will be at its maximum when the part being subtracted from the 2 is as small as it can be. That part after the minus sign, $(x + 2)^2$, is the square of something, so it can be no smaller than 0. When $x = -2$, $(x + 2)^2 = 0$, and the whole expression $2 - (x + 2)^2 = 2 - 0 = 2$. For any other value of x, the part after the minus sign will be greater than 0, and the whole expression will be less than 2. So 2 is the maximum value, and the answer is (D).

 Watch Those Parentheses.

When it comes to functions, pay close attention to order and parentheses.

 Compound Functions

$f(g(x))$ means apply g first, and then apply f to the result.

$g(f(x))$ means apply f first, and then apply g to the result.

$f(x)g(x)$ means apply f and g separately, and then multiply the results.

 Maximums and Minimums

To find a maximum or minimum value of a function, look for parts of the expression—especially squares—that have upper or lower limits.

Graphing Functions

Like just about everything else with functions, graphing is no big deal once you understand the conventions. Example 4 provides a very good illustration.

Example 4

If $f(x) = |1 - x|$, which of the following could be the graph of $y = f(x)$?

(A)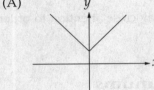

(B)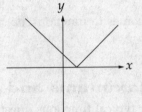

(C)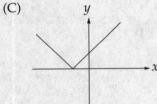

(D)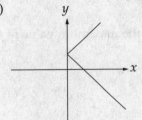

(E)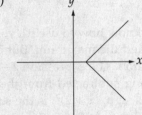

The question says $f(x) = |1 - x|$ and asks for the graph of $y = f(x)$. Well, that just means the graph of $y = |1 - x|$. The smallest that y can be is 0, because the absolute value of anything is nonnegative. That narrows the choices to (B) and (C). Notice that (D) and (E) cannot be functions, because for some values of x they show *two* associated values of y. A function, by definition, yields no more than one y for any particular x.

To choose between (B) and (C), think about what value of x yields $y = 0$.

$$|1 - x| = 0$$
$$1 - x = 0$$
$$-x = -1$$
$$x = 1$$

So the graph touches the x-axis on the *positive* side, and the answer is (B).

★ Level IIC Topics

The rest of the topics in this chapter are tested primarily on the SAT II: Mathematics Level IIC Test. If you're preparing for Level IC, you're finished with your review of functions. To see how much you've learned, try the first four questions in the Follow-Up Test at the end of the chapter. If you're preparing for the SAT II: Mathematics Level IIC Test, you'll need to read on. The next two topics are for you.

Undefined

The issue of "defined" and "undefined" functions is brought up by Example 5. This question is typical of what you might encounter on the Level IIC Test.

★Example 5

If $f(x) = \dfrac{\sqrt{x^2 - 4}}{x - 4}$, what are all the values of x for which $f(x)$ is defined?

(A) All real numbers except 4
(B) All real numbers except −2 and 2
(C) All real numbers greater than or equal to −2 and less than or equal to 2
(D) All real numbers less than or equal to −2 or greater than or equal to 2
(E) All real numbers less than or equal to −2 or greater than or equal to 2, except 4

There are two things to watch out for when looking for values for which a function is undefined. First, watch out for division by zero. And second, watch out for negatives under radicals. In this case you have both.

Find Your Next Move

The rest of the topics in this section relate primarily to SAT II: Mathematics Level IIC. If you're preparing for Level IC, skip ahead to the Follow-Up Test at the end of the chapter.

Undefined

To find values of x for which a function is undefined, look for values that would make a denominator zero or that would make an expression under a radical negative.

The function is undefined when the denominator is zero:

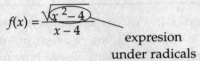

$$f(x) = \frac{\sqrt{x^2 - 4}}{x - 4}$$

denominator

$$x - 4 = 0$$
$$x = 4$$

The function is also undefined when the expression under the radical is negative:

$$f(x) = \frac{\sqrt{x^2 - 4}}{x - 4}$$

expresion
under radicals

$$x^2 - 4 < 0$$
$$-2 < x < 2$$

So the expression is undefined when $x = 4$ or when $-2 < x < 2$, and the answer is (E).

 ## Inverse Functions

Perhaps the most advanced functions question you'll face on the SAT II: Level IIC will be one involving the inverse of a function—written $f^{-1}(x)$ —as in Example 6.

★**Example 6**

If $f(x) = \frac{1}{3}x + 3$, then $f^{-1}(x) =$

(A) $-\frac{1}{3}x - 3$

(B) $-3x + \frac{1}{3}$

(C) $3x + \frac{1}{3}$

(D) $3x - 3$

(E) $3x - 9$

Here's what you do to find the inverse of a function. First, put y in the place of $f(x)$ to make a more familiar-looking equation form:

$$f(x) = \frac{1}{3}x + 3$$
$$y = \frac{1}{3}x + 3$$

Second, solve the equation for x in terms of y :

$$y = \frac{1}{3}x + 3$$

$$y - 3 = \frac{1}{3}x$$

$$3(y - 3) = x$$

$$x = 3y - 9$$

Third, put $f^{-1}(x)$ in the place of x and put x in the place of y :

$$x = 3y - 9$$

$$f^{-1}(x) = 3x - 9$$

The answer is (E).

Now that you've had a good look at some typical SAT II: Mathematics functions questions, it's time to try a few more on your own. See how well you can do with the questions in the Functions Follow-Up Test.

Inverse Functions

- To find the inverse of a function:
  1. Replace $f(x)$ with y.
  2. Solve the equation for x in terms of y.
  3. Replace x with $f^{-1}(x)$ and replace y with x.

- Graphs of inverse functions are symmetric about the line $y = x$.

- The slopes of the lines of inverse functions are reciprocals.

Functions Follow-Up Test

Level IC: 4 Questions (6 Minutes)
Level IIC: 6 Questions (8 Minutes)

Directions for Level IC: If you are preparing for the SAT II: Mathematics Level IC Test, solve problems 1–4 and choose the best answer from those given. Fill in the oval corresponding to the best answer choice in the grid to the right of each question. (Answers and explanations begin on page 180.)

Directions for Level IIC: If you are preparing for the SAT II: Mathematics Level IIC Test, solve problems 1–6 and choose the best answer from those given. Fill in the oval corresponding to the best answer choice in the grid to the right of each question. (Answers and explanations begin on page 180.)

DO YOUR FIGURING HERE.

1. If $f(x) = x^2 + x$ for all x, and if $f(a - 1) = -\dfrac{1}{4}$, what is the value of a ?

(A) $-\dfrac{1}{2}$

(B) $-\dfrac{1}{4}$

(C) $\dfrac{1}{4}$

(D) $\dfrac{1}{2}$

(E) $\dfrac{3}{4}$ Ⓐ Ⓑ Ⓒ Ⓓ Ⓔ

2. If $f(x) = x^2$ and $g(x) = 2x$, what is the value of $f(g(-3)) - g(f(-3))$?

(A)   54

(B)   18

(C)   0

(D)   −18

(E)   −54   Ⓐ Ⓑ Ⓒ Ⓓ Ⓔ

3. For what value of x is $f(x) = 7 - (4 - x)^2$ at its maximum?

DO YOUR FIGURING HERE.

(A)   –9

(B)   –3

(C)   3

(D)   4

(E)   7   Ⓐ Ⓑ Ⓒ Ⓓ Ⓔ

4. The graph in Figure 1 could be the graph of which of the following functions?

(A) $f(x) = (x - 2)^2$

(B) $f(x) = x^2 - 4$

(C) $f(x) = 4 - x^2$

(D) $f(x) = |x^2 - 4|$

(E) $f(x) = -|x^2 - 4|$ Ⓐ Ⓑ Ⓒ Ⓓ Ⓔ

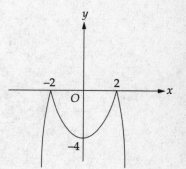

Figure 1

★5. If $f(x) = \dfrac{1}{\sqrt{1 - x^2}}$, which of the following describes all the real values of x for which $f(x)$ is undefined?

(A) $x = -1$ or $x = 1$

(B) $x < -1$ or $x > 1$

(C) $x \le -1$ or $x \ge 1$

(D) $-1 < x < 1$

(E) $-1 \le x \le 1$ Ⓐ Ⓑ Ⓒ Ⓓ Ⓔ

DO YOUR FIGURING HERE.

★6. If $f(x)$ is a linear function and the slope of $y = f(x)$ is $\frac{1}{2}$, what is the slope of $y = f^{-1}(x)$?

(A) -2

(B) $-\dfrac{1}{2}$

(C) $\dfrac{1}{2}$

(D) 2

(E)   It cannot be determined from the information given.

Ⓐ Ⓑ Ⓒ Ⓓ Ⓔ

STOP!   **END OF TEST. DO NOT TURN THE PAGE UNTIL YOU ARE READY TO CHECK YOUR ANSWERS.**

Turn the page
for answers and explanations
to the Follow-Up Test.

Follow-Up Test—Answers and Explanations

Answer Key 1. *D*   2. *B*   3. *D*   4. *E*   5. *C*   6. *D*

1. **(D)**—Plug $x = a - 1$ into the definition of the function:

$$f(x) = x^2 + x$$
$$f(a-1) = (a-1)^2 + (a-1)$$
$$= a^2 - 2a + 1 + a - 1$$
$$= a^2 - a$$

Now write an equation in which the result is equal to $-\dfrac{1}{4}$:

$$a^2 - a = -\frac{1}{4}$$
$$a^2 - a + \frac{1}{4} = 0$$
$$4a^2 - 4a + 1 = 0$$
$$(2a - 1)^2 = 0$$
$$2a - 1 = 0$$
$$2a = 1$$
$$a = \frac{1}{2}$$

2. **(B)**—Perform the inside functions first:

$$g(x) = 2x$$
$$g(-3) = 2(-3) = -6$$

$$f(x) = x^2$$
$$f(-3) = (-3)^2 = 9$$

Then perform the outside functions on the results:

$$f(x) = x^2$$
$$f(g(-3)) = f(-6) = (-6)^2 = 36$$

$$g(x) = 2x$$
$$g(f(-3)) = g(9) = 2(9) = 18$$

Now you can find the difference:

$$f(g(-3)) - g(f(-3)) = 36 - 18 = 18$$

3. **(D)**—The expression $7 - (4 - x)^2$ will be at its maximum when the part after the minus sign, $(4 - x)^2$, is as small as it can be. Something squared can never be less than zero, so the whole expression will be at its maximum when $(4 - x)^2 = 0$:

$$(4 - x)^2 = 0$$
$$4 - x = 0$$
$$x = 4$$

4. **(E)**—The graph shows three points that represent solutions to the equation $y = f(x)$: $(-2, 0)$, $(2, 0)$, and $(0, -4)$. The first point satisfies choices (B), (C), (D) and (E). The second point satisfies all the answer choices. The point $(0, -4)$, however, satisfies only choices (B) and (E). How can you decide between those choices? Think about some other point on the graph. What about when $x = 3$? Then, according to choice (B), y would be $(3)^2 - 4 = 5$; while according to (E), y would be $-\left|(3)^2 - 4\right| = -5$. You can see from the graph that when $x = 3$, y is negative, so the only choice that fits is (E).

★5. **(C)**—The function will be undefined for any values of x that make the denominator zero or that make the expression under the radical negative. The denominator is zero when:

$$\sqrt{1 - x^2} = 0$$
$$1 - x^2 = 0$$
$$x^2 = 1$$
$$x = \pm 1$$

The expression under the radical is negative when:

$$1 - x^2 < 0$$
$$-x^2 < -1$$
$$x^2 > 1$$
$$x < -1 \text{ or } x > 1$$

So the function is undefined when $x \le -1$ or when $x \ge 1$.

★6. **(D)**—You might think this one can't be done because it does not tell you exactly what the function is that you want the inverse of. It just says the slope is $\frac{1}{2}$. But, in fact, it doesn't actually ask for the inverse function; it just asks for the slope. You can spot the correct answer immediately if you remember how the slopes of inverse functions are related: they're reciprocals. But you can still find the answer without remembering that relationship. You know the slope of $f(x)$ is $\frac{1}{2}$. You don't know the y-intercept, so just call it b:

$$f(x) = \frac{1}{2}x + b$$

Find the inverse function:

$$f(x) = \frac{1}{2}x + b$$
$$y = \frac{1}{2}x + b$$
$$y - b = \frac{1}{2}x$$
$$2y - 2b = x$$
$$x = 2y - 2b$$
$$f^{-1}(x) = 2x - 2b$$

The slope is the coefficient of x, which is 2.

What's Next

The next chapter is a roundup of a variety of topics that are likely to appear on the SAT II: Mathematics Subject Test.

CHAPTER 10

Miscellaneous Topics

This chapter covers a dozen largely unrelated math topics. What these diverse topics have in common is that they are all regularly tested on the SAT II: Mathematics Subject Test. You will not see all the topics in this chapter on any one edition of the Mathematics Level IC or Level IIC Test, but you can be sure that *some* of these topics will appear.

How to Use This Chapter

Maybe you already know everything you need to know about these dozen math topics. You can find out by taking the Miscellaneous Topics Diagnostic Test on page 184. The first seven questions on the Diagnostic Test are typical of what you could expect on either a Level IC or a Level IIC Mathematics Subject Test. Questions 8–12, though, would appear only on a Level IIC Test. We've marked those Level IIC questions with stars. Check your answers using the answer key following the test. No matter how you score, don't worry! The answer key also shows where to find a detailed explanation for each question. The "Find Your Study Plan" section that follows the test will suggest next steps based on your performance on the Diagnostic.

Find Your Level

How you use this chapter really depends on which test you're taking and how much time you have to prep. Find your level and pace below.

Taking the Mathematics Level IC Test? Try the first seven questions in the Miscellaneous Topics Diagnostic Test. Skip the Level IIC questions that we've marked with a star. Read the rest of the chapter, skipping any material that we've flagged as relating to Level IIC only. Then try the first seven questions in the Follow-Up Test at the end of the chapter.

Miscellaneous Facts and Formulas in This Chapter

- Rules of Imaginary Numbers (p. 191)
- Contrapositive (p. 192)
- Percent Increase and Decrease Formulas (p. 195)
- Average, Median, and Mode (p. 196)
- Average Rate Formula (p. 197)

Topics Relating to Level IIC only:

- Permutations and Combinations Formulas (p. 198)★
- Probability Formulas (p. 199)★
- Rules of Logarithms (p. 200)★
- Limit of an Algebraic Fraction (p. 201)★
- Sequences Formulas (p. 202)★

★
What's This Mean?

This icon appears next to more difficult topics and questions that would only appear on the Level IIC Test.

Kaplan Strategies in This Chapter

- Don't freak out (p. 189).
- Be suspicious (p. 194).
- Use the sum (p. 196).
- Avoid the speed trap (p. 197).

Mathematics Level IC Shortcut Try the first seven questions in the Miscellaneous Topics Diagnostic Test. The "Find Your Study Plan" section that follows the test will suggest next steps based on your Diagnostic Test score.

Taking the Mathematics Level IIC Test? Do everything in this chapter. It's all relevant to the Level IIC Test.

Mathematics Level IIC Shortcut Take the Miscellaneous Topics Diagnostic Test and check your answers. The "Find Your Study Plan" section that follows the test will suggest next steps based on your Diagnostic Test score.

Panic Plan? Make sure you're comfortable with the fundamental material in previous chapters before you spend any time on these miscellaneous topics.

Miscellaneous Topics Diagnostic Test

Level IC: 7 Questions (9 Minutes)
Level IIC: 12 Questions (15 Minutes)

Directions for Level IC: If you are preparing for the SAT II: Mathematics Level IC Test, solve problems 1–7 and choose the best answer from those given. Fill in the oval corresponding to the best answer choice in the grid to the right of each question. (Answers are on page 190.)

Directions for Level IIC: If you are preparing for the SAT II: Mathematics Level IIC Test, solve problems 1–12 and choose the best answer from those given. Fill in the oval corresponding to the best answer choice in the grid to the right of each question. (Answers are on page 190.)

DO YOUR FIGURING HERE.

1. An operation § is defined for all real numbers a and b by the equation $a \mathbin{\S} b = \dfrac{a^2}{2} + \dfrac{b}{3}$. If $-3 \mathbin{\S} x = 0$, what is the value of x?

    (A) -13.5

    (B) -9

    (C) 0

    (D) 3

    (E) 9

DO YOUR FIGURING HERE.

2. If $i^2 = -1$, then $(2 - 3i)^2 =$

    (A) $-5 - 12i$

    (B) $4 + 6i$

    (C) $4 + 12i$

    (D) $5 + 12i$

    (E) $13 - 6i$

    Ⓐ Ⓑ Ⓒ Ⓓ Ⓔ

3. "If p, then q," is logically equivalent to which of the following?
    I.   If q, then p.
    II. If not p, then not q.
    III. If not q, then not p.

    (A) I only

    (B) III only

    (C) I and II only

    (D) I and III only

    (E) I, II, and III

    Ⓐ Ⓑ Ⓒ Ⓓ Ⓔ

4. The charge for a taxi ride in a certain city is 2 dollars for the first one-fifth of a mile and 20 cents for each additional one-fifth of a mile. Which of the following equations represents the charge, in cents, for a taxi ride of exactly x miles, where x is a positive integer?

    (A) $2x + 80$

    (B) $3x$

    (C) $100x + 180$

    (D) $200x + 80$

    (E) $300x$

    Ⓐ Ⓑ Ⓒ Ⓓ Ⓔ

DO YOUR FIGURING HERE.

5. During the eighteenth century, the population of country X increased by 1,000 percent. During the nineteenth century, the population increased by 500 percent. By what percent did the population increase over the two-century period?

    (A) 1,500%

    (B) 1,700%

    (C) 4,900%

    (D) 5,000%

    (E) 6,500%

6. George's average (arithmetic mean) score on the first four English tests of the semester is 92. If he earns a score of 77 on the fifth test, what will his new average be?

    (A) 84.5

    (B) 86.0

    (C) 87.5

    (D) 88.0

    (E) 89.0

7. Martha drives half the distance from A to B at 40 miles per hour and the other half the distance at 60 miles per hour. What is her average rate of speed, in miles per hour, for the entire trip?

    (A) 48

    (B) 49

    (C) 50

    (D) 51

    (E) 52

DO YOUR FIGURING HERE.

*8. A committee of 3 women and 2 men is to be formed from a pool of 11 women and 7 men. How many different committees can be formed?

(A) 3,465

(B) 6,930

(C) 10,395

(D) 20,790

(E) 41,580 Ⓐ Ⓑ Ⓒ Ⓓ Ⓔ

*9. The probability of rain today is 60 percent, and the independent probability of rain tomorrow is 75 percent. What is the percent probability that it will rain neither today nor tomorrow?

(A) 10%

(B) 15%

(C) 45%

(D) 55%

(E) 65% Ⓐ Ⓑ Ⓒ Ⓓ Ⓔ

*10. $\log_3 \sqrt[9]{3} =$

(A) $\dfrac{1}{9}$

(B) $\dfrac{1}{3}$

(C) $\dfrac{1}{2}$

(D) 1

(E) 3 Ⓐ Ⓑ Ⓒ Ⓓ Ⓔ

DO YOUR FIGURING HERE.

★11. $\lim\limits_{x \to 3} \dfrac{x^3 + x^2 - 12x}{x^2 - 9} =$

(A)   –7

(B) $-\dfrac{7}{2}$

(C)   0

(D) $\dfrac{7}{2}$

(E)   The limit does not exist.   Ⓐ Ⓑ Ⓒ Ⓓ Ⓔ

★12. If the first term in a geometric sequence is 3, and if the third term is 48, what is the 11th term?

(A)   228

(B)   528

(C)   110,592

(D)   3,145,728

(E)   12,582,912   Ⓐ Ⓑ Ⓒ Ⓓ Ⓔ

STOP!   **END OF TEST. DO NOT TURN THE PAGE UNTIL YOU ARE READY TO CHECK YOUR ANSWERS.**

Find Your Study Plan

The answer key is on page 190. It doesn't matter so much how you did on the Miscellaneous Topics Diagnostic Test as a whole, because the topics are so varied. What matters is exactly what questions you had trouble with. No matter what your Diagnostic Test score is, or what level test you're preparing for, your next step is to read and study those parts of this chapter that address the particular diagnostic questions you got wrong. The answer key shows where in this chapter to find explanations for the questions you missed.

Test Topics

This icon appears next to each discussion of a math topic that's tested on the SAT II.

Miscellaneous Test Topics

We'll use the questions on the Miscellaneous Topics Diagnostic Test to point out a dozen more topics you'll need to be prepared to deal with on the SAT II: Mathematics Subject Test. Only the first seven topics are relevant to the Mathematics Level IC test. If you're preparing for the Level IC Test, you can skip the discussion of the last five topics. If you're preparing for Level IIC, however, this entire chapter will be relevant to you.

Don't Freak Out.

If you see something—a symbol, a word—you've never seen before, chances are the test makers just made it up to test your ability to stay cool in the face of something new and unfamiliar.

 Symbolism, Definitions, and Instructions

One miscellaneous question type that turns up with great regularity is the type that presents you with a new symbol representing an unfamiliar operation, or a new word with an unfamiliar definition, or a new procedure with unfamiliar steps. In Example 1 you have a new symbol and operation:

Example 1

An operation § is defined for all real numbers a and b by the equation $a \, \S \, b = \dfrac{a^2}{2} + \dfrac{b}{3}$. If $-3 \, \S \, x = 0$, what is the value of x ?

(A) -13.5

(B) -9

(C) 0

(D) 3

(E) 9

Diagnostic Test Answers and Reviews

1.   A
See "Symbolism, Definitions, and Instructions," p. 189.

2.   A
See "Imaginary and Complex Numbers," p. 190.

3.   B
See "Logic," p. 192.

4.   C
See "Translating from English into Algebra," p. 193.

5.   E
See "Percent Increase and Decrease," p. 194.

6.   E
See "Averages," p. 195.

7.   A
See "Distance, Rate, and Time," p. 196.

8.   A
See "Permutations and Combinations," p. 198.*

9.   A
See "Probability," p. 199.*

10.   A
See "Logarithms," p. 200.*

11.   D
See "Limits," p. 201.*

12.   D
See "Sequences," p. 202.*

The very point of a question like this is to test your ability not to freak out at the sight of something new and unfamiliar. You're not expected to have any prior experience with this operation: The test makers just made it up! But they also defined it for you, so as long as you don't freak out, everything you need to answer the question is right there.

The operation § is defined as $a § b = \dfrac{a^2}{2} + \dfrac{b}{3}$, and when $a = -3$ and $b = x$, the result is 0.

$$a§b = \dfrac{a^2}{2} + \dfrac{b}{3}$$

$$3§x = \dfrac{3^2}{2} + \dfrac{x}{3} = 0$$

$$\dfrac{9}{2} + \dfrac{x}{3} = 0$$

$$\dfrac{x}{3} = -\dfrac{9}{2}$$

$$x = -\dfrac{27}{2} = -13.5$$

The answer is (A).

Imaginary and Complex Numbers

There's a chance you will encounter a question that begins, "If $i^2 = -1$ . . . ," like the example below:

Example 2

If $i^2 = -1$, then $(2 - 3i)^2 =$

(A) $-5 - 12i$

(B) $4 + 6i$

(C) $4 + 12i$

(D) $5 + 12i$

(E) $13 - 6i$

Imaginary numbers: To answer an imaginary numbers question like this, you have to know how to use your i's. When you're adding or subtracting, i's act a lot like variables:

$$i + i = 2i$$
$$i - i = 0$$
$$2i + 3i = 5i$$
$$2i - 3i = -i$$

But when you're multiplying, or raising to a power, don't forget to take that extra step of changing i^2 to -1:

$$i \times i = i^2 = -1$$
$$(2i)(3i) = 6i^2 = 6(-1) = -6$$
$$(3i)^2 = (3i)(3i) = 9i^2 = 9(-1) = -9$$

Notice that when you raise i to successive powers, a pattern develops:

$$i^1 = i \qquad i^5 = i \qquad i^9 = i$$
$$i^2 = -1 \qquad i^6 = -1 \qquad i^{10} = -1$$
$$i^3 = -i \qquad i^7 = -i \qquad i^{11} = -i$$
$$i^4 = 1 \qquad i^8 = 1 \qquad i^{12} = 1$$

Complex number: A number in the form $a + bi$, where a and b are real numbers, is called a *complex number*. In Example 2, you're looking for the square of a complex number. To multiply complex numbers, use FOIL. (Review Chapter 4 if you forgot what FOIL stands for. It's a way of remembering the order of operations for multiplying binomials: first, outer, inner, last.) Just remember again to take that extra step of changing i^2 to -1:

$$(2 - 3i)(2 - 3i)$$
$$= (2)(2) + (2)(-3i) + (-3i)(2) + (-3i)(-3i)$$
$$= 4 - 6i - 6i + 9i^2$$
$$= 4 - 12i - 9$$
$$= -5 - 12i$$

The answer is (A).

Rules of Imaginary Numbers

$$ai + bi = (a + b)i$$
$$ai - bi = (a - b)i$$
$$(ai)(bi) = -ab$$
$$i^2 = -1$$
$$i^3 = -i$$
$$i^4 = 1$$

Contrapositive

"If *p*, then *q*," is logically
equivalent to "If not *q*, then
not *p*."

 Logic

If you encounter a logic question on an SAT II: Mathematics Test, it is
likely that about all you'll have to know is the so-called *contrapositive*.
That's what Example 3 is getting at.

Example 3

"If *p*, then *q*," is logically equivalent to which of the following?
    I.   If *q*, then *p*.
    II. If not *p*, then not *q*.
    III. If not *q*, then not *p*.

(A)   I only

(B)   III only

(C)   I and II only

(D)   I and III only

(E)   I, II, and III

The original statement "if *p*, then *q*" is a general form that covers such statements as the following:

1. If you live in Alabama, then you live in the United States (*p* = "live
   in Alabama"; *q* = "live in U.S.").
2. All prime numbers are integers (*p* = "is prime"; *q* = "is an integer").
3. If Marla studies, she will get an A on the test (*p* = "studies"; *q* = "gets
   an A").

Of the three Roman numeral options in Example 3, only one necessarily follows. Take a look at the options one at a time:

I. If *q*, then *p*. This is *not necessarily so.* You cannot simply switch the *p*
   and the *q*. Look what illogical results you would get with the three
   samples above:

1. If you live in the United States, then you live in Alabama.
2. All integers are prime numbers.
3. If Marla gets an A on the test, then she must have studied.

Can you see that none of these statements—not even number 3—necessarily follows from the original?

II. If not *p*, then not *q*. This is *not necessarily so.* You cannot simply negate
    both the *p* and the *q*. Look what illogical results you would get with
    the three samples above:

1. If you don't live in Alabama, then you don't live in the United States.
2. If a number's not prime, then it's not an integer.
3. If Marla doesn't study, then she won't get an A.

Can you see that none of these statements necessarily follows from the original statement?

III. If not *q*, then not *p*. This is *true*. If you both switch the *p* and *q* *and* negate them, the result is logically equivalent to the original. This is the *contrapositive*. Here are the contrapositives of the three samples above:

1. If you don't live in the United States, then you don't live in Alabama.
2. If a number's not an integer, then it's not a prime number.
3. If Marla doesn't get an A on the test, then she must not have studied.

These three statements are all as true as the originals they're based on. So, of the three options, only III is true, and the answer is (B).

 Translating from English into Algebra

Solving a word problem means taking a situation that is described verbally and turning it into one that is described mathematically. It means translating from English into algebra, which is exactly what Example 4 asks you to do.

Example 4

The charge for a taxi ride in a certain city is 2 dollars for the first one-fifth of a mile and 20 cents for each additional one-fifth of a mile. Which of the following equations represents the charge, in cents, for a taxi ride of exactly *x* miles, where *x* is a positive integer?

(A) $2x + 80$

(B) $3x$

(C) $100x + 180$

(D) $200x + 80$

(E) $300x$

To translate successfully, you must first be sure you understand the original. It's usually a good idea to read a word problem question stem twice. The charge for the first fifth of a mile is 2 dollars. You want your answer in cents, so write:

$$\text{charge} = 200 + \cdots$$

> **"It's usually a good idea to read a word-problem question twice."**

No matter how far you go, your fare starts with 200 cents. On top of that you pay 20 cents for each *additional* fifth of a mile:

$$\text{charge} = 200 + 20(\text{\# of additional fifth-miles})$$

The trickiest part of this translation is figuring out how to express the number of additional fifth-miles in terms of x.

If x is the number of whole miles, and each whole mile contains 5 fifth-miles, the total number of fifth-miles is $5x$. But you don't have to pay 20 cents for every one of those $5x$ fifth-miles—you already paid for the first fifth-mile. The number of fifth-miles after the first is $5x - 1$:

$$\text{charge} = 200 + 20(5x - 1) = 200 + 100x - 20 = 100x + 180$$

The answer is (C).

 ## Percent Increase and Decrease

Some word problems present classic situations for which there are standard approaches, or even formulas. Example 5, for instance, requires that you know how to find a percent increase:

Example 5

During the eighteenth century, the population of country X increased by 1,000 percent. During the nineteenth century, the population increased by 500 percent. By what percent did the population increase over the two-century period?

(A) 1,500%

(B) 1,700%

(C) 4,900%

(D) 5,000%

(E) 6,500%

First, you should realize that answering this question is going to take more than merely adding 1,000 percent and 500 percent to get (A) 1,500 percent. You're not going to find a word problem on the SAT II for which all you have to do is add two numbers.

Be Suspicious.

If the way to solve a word problem seems obvious and simple, think again. The problem is probably more complicated than it looks at first glance.

This word problem is extra complicated for two reasons. First, the percents are so huge. You don't often have to work with "1,000 percent" or "500 percent." Second, you have a compound percent increase situation, a situation that's always trickier than it looks at first.

Call the population of country X at the beginning of the eighteenth century P. That number increases by 1,000 percent. That is, it goes up by 1,000 percent of itself. One thousand percent of P means $10P$, so if P goes up by that much, it becomes $P + 10P = 11P$.

$$P + (1,000\% \text{ of } P) = P + 10P = 11P$$

Now the population at the beginning of the nineteenth century is $11P$. That number increases by 500 percent. In other words, on top of $11P$ you add 500 percent of $11P$:

$$11P + (500\% \text{ of } 11P) = 11P + 5(11P) = 66P$$

The population at the end of the nineteenth century is $66P$. That's $65P$ more than the original population P:

$$66P = P + 65P$$
$$= P + (6,500\% \text{ of } P)$$

The net increase is 6,500 percent, and the answer is (E).

Averages

Another classic situation with a standard approach is an averages question like Example 6.

Example 6

George's average (arithmetic mean) score on the first four English tests of the semester is 92. If he earns a score of 77 on the fifth test, what will his new average be?

(A) 84.5

(B) 86.0

(C) 87.5

(D) 88.0

(E) 89.0

Percent Increase and Decrease Formulas

Percent increase =

$$\frac{\text{Amount of increase}}{\text{Original Amount}} \times 100\%$$

Percent decrease =

$$\frac{\text{Amount of decrease}}{\text{Original Amount}} \times 100\%$$

Average, Median, and Mode

Average (arithmetic mean)
$$= \frac{\text{Sum of the terms}}{\text{Number of terms}}$$

Median = middle value
(or average of two middle values)

Mode = most frequent value

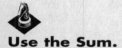

Use the Sum.

The key to many averages questions is to use this variant of the Average formula:

Sum of the terms = (Average) × (Number of terms)

The key to almost every SAT II averages question is to think about the sum. Everybody knows the standard averages formula:

$$\text{Average} = \frac{\text{Sum of the terms}}{\text{Number of terms}}$$

Perhaps less familiar, but often more useful, is this other version of the averages formula:

$$\text{Sum of the terms} = (\text{Average}) \times (\text{Number of terms})$$

George's average after four tests is 92. To find out what effect a 77 on test number 5 will have, calculate the sum of the first four scores:

$$\text{Sum} = \text{Average} \times \text{Number} = 92 \times 4 = 368$$

To get the new average, add 77 and divide by 5:

$$\text{New Average} = \frac{368 + 77}{5} = \frac{445}{5} = 89.0$$

The answer is (E).

 ## Distance, Rate, and Time

Yet another classic situation with a standard approach is the distance-rate-and-time question. Example 7 is an interesting variation.

Example 7

Martha drives half the distance from *A* to *B* at 40 miles per hour and the other half the distance at 60 miles per hour. What is her average rate of speed, in miles per hour, for the entire trip?

(A) 48

(B) 49

(C) 50

(D) 51

(E) 52

At first you might think, "Halfway at 40 mph, halfway at 60 mph, the average is 50." But that's way too simple for an SAT II word problem. The average rate would be 50 if Martha drove half the time at 40 and half the time at 60. But, it's half the distance she drives at 40 and half the distance she drives at 60. So she actually spends more time driving at the slower speed, and so her average speed will be something closer to 40 than to 60. So the answer's going to be either (A) or (B).

Average speed is defined as **total distance divided by total time:**

$$\text{Average speed} = \frac{\text{Total distance}}{\text{Total time}}$$

In Example 7 the distance is not specified, so call it d. To get the total time, think about the time for each half-distance. Martha travels the first $\frac{d}{2}$ miles at 40 miles per hour:

$$\text{Time}_1 = \frac{\text{Distance}}{\text{Rate}} = \frac{\frac{d}{2}}{40} = \frac{d}{80}$$

She travels the second $\frac{d}{2}$ miles at 60 miles per hour:

$$\text{Time}_2 = \frac{\text{Distance}}{\text{Rate}} = \frac{\frac{d}{2}}{60} = \frac{d}{120}$$

The total time, then, is:

$$\text{Total time} = \frac{d}{80} + \frac{d}{120}$$
$$= \frac{3d}{240} + \frac{2d}{240} = \frac{5d}{240} = \frac{d}{48}$$

Now plug Total distance = d and Total time = $\frac{d}{48}$ into the general average speed formula:

$$\text{Average speed} = \frac{\text{Total distance}}{\text{Total time}} = \frac{d}{\frac{d}{48}} = 48$$

The answer is (A).

★ Level IIC Topics

The rest of the topics in this chapter are tested only on the SAT II: Mathematics Level IIC Test. If you're preparing for Level IC, congratulations! You're finished with your review of SAT II: Mathematics topics. To see how much you've learned, try the first seven questions in the Follow-Up Test at the end of the chapter. If you're preparing for the Level IIC Test, you'll need to read on. The next five topics are for you.

Avoid the Speed Trap.

When a problem asks for average rate or average speed, don't just average the given rates or speeds. It's not that simple. Use the Average Rate formula.

Average Rate Formula

Average A per B
$$= \frac{\text{Total } A}{\text{Total } B}$$

Average speed
$$= \frac{\text{Total distance}}{\text{Total time}}$$

Find Your Next Move

The rest of the topics in this chapter relate to SAT II: Mathematics Level IIC only. If you're preparing for Level IC, skip ahead to the Follow-Up Test at the end of the chapter.

Permutations and Combinations

To be successful with combinations and permutations questions like Example 8, you have to remember the relevant formulas.

Permutations and Combinations Formulas

The number of permutations of n distinct objects is:
$$n! = n(n-1)(n-2)\cdots(3)(2)(1)$$

The number of permutations of n objects, a of which are indistinguishable, and b of which are indistinguishable, etc., is:

$$\frac{n!}{a!b!\ldots}$$

The number of permutations of n objects taken r at a time, is:

$$_nP_r = \frac{n!}{(n-r)!}$$

The number of combinations of n objects taken r at a time, is:

$$_nC_r = \frac{n!}{r!(n-r)!}$$

*Example 8

A committee of 3 women and 2 men is to be formed from a pool of 11 women and 7 men. How many different committees can be formed?

(A)   3,465

(B)   6,930

(C)   10,395

(D)   20,790

(E)   41,580

This question asks about the number of groups that can be formed, so it's a *combinations* question. The number of combinations of 11 women taken 3 at a time is:

$$_{11}C_3 = \frac{11!}{3!(11-3)!}$$

$$= \frac{11!}{3!(8!)}$$

$$= \frac{11 \times 10 \times 9}{3 \times 2}$$

$$= 165$$

The number of combinations of 7 men taken 2 at a time is:

$$_7C_2 = \frac{7!}{2!(7-2)!}$$

$$= \frac{7!}{2!(5!)}$$

$$= \frac{7 \times 6}{2}$$

$$= 21$$

For each of the 165 combinations of women there are 21 combinations of men, so the combined number of combinations is $165 \times 21 = 3,465$. The answer is (A).

If you're good at memorizing formulas—if you can master the permutations and combinations formulas without too much trouble—go ahead and do it. But all these formulas probably won't get you more than one right answer on the Level II test, so they don't deserve a whole lot of time and effort.

 ## Probability

To answer an SAT II probability question, you need to know not only how to use the general probability formula, but also how to deal with multiple probabilities and probabilities that events will *not* occur. You have a little bit of everything in Example 9.

*Example 9

The probability of rain today is 60 percent, and the independent probability of rain tomorrow is 75 percent. What is the percent probability that it will rain neither today nor tomorrow?

(A) 10%

(B) 15%

(C) 45%

(D) 55%

(E) 65%

To get the probability that an event will *not* occur, subtract the probability that it *will* occur from 1. The probability that it will rain today is 60%, so the probability that it will not rain today is 100% − 60% = 40%. The probability that it will rain tomorrow is 75%, so the probability that it will not rain tomorrow is 100% − 75% = 25%.

To get the probability that two independent events will both occur, multiply the individual probabilities. Thus, to get the probability that it will not rain today or tomorrow, multiply 40% and 25%:

$$(40\%)(25\%) = (.40)(.25) = .10 = 10\%$$

The answer is (A).

Probability Formulas

Probability =
$\dfrac{\text{\# of favorable outcomes}}{\text{Total \# of possible outcomes}}$

If the probability that an event will occur is a, then the probability that the event will not occur is $1 - a$.

If the probability that one event will occur is a and the independent probability that another event will occur is b, then the probability that both events will occur is ab.

Logarithms

Maybe you'll see a logarithms question like Example 10 on your test.

★**Example 10**

$$\log_3 \sqrt[9]{3} =$$

(A) $\dfrac{1}{9}$

(B) $\dfrac{1}{3}$

(C) $\dfrac{1}{2}$

(D) 1

(E) 3

Rules of Logarithms

$$\log_b(XY) = \log_b X + \log_b Y$$

$$\log_b\left(\frac{X}{Y}\right) = \log_b X - \log_b Y$$

$$\log_b X^Y = Y\log_b X$$

$$\log_b b = 1$$

$$\log_b X = \frac{\log_a X}{\log_a b}$$

Logarithms are not hard once you're familiar and experienced with them and know all the rules. If you're a long way from fully understanding logs, don't worry about them for this test. Logs will probably account for no more than one question. It's not worth spending a lot of time and effort figuring out logs when they appear so infrequently on the test.

If you understand what a log means, then Example 10 is not very difficult. The log to the base 3 of $\sqrt[9]{3}$ is the exponent that attached to the base 3 will give you $\sqrt[9]{3}$:

$$\log_3 \sqrt[9]{3} = x$$

$$3^x = \sqrt[9]{3}$$

Now what you're doing is solving an equation with the variable in an exponent. The way to do that is to reexpress both sides of the equation in terms of the same base.

$$3^x = \sqrt[9]{3}$$

$$3^x = 3^{\frac{1}{9}}$$

$$x = \frac{1}{9}$$

The answer is (A).

Learn the rules of logarithms if you want to be sure to get a possible logarithms question right. But don't be afraid to ignore logs and to skip the logs question if you get one.

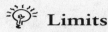 **Limits**

If you see a limits question like Example 11, forget about abstract theorems and calculus. There's a standard approach to a question like Example 11.

★**Example 11**

$$\lim_{x \to 3} \frac{x^3 + x^2 - 12x}{x^2 - 9} =$$

(A) −7

(B) $-\dfrac{7}{2}$

(C) 0

(D) $\dfrac{7}{2}$

(E) The limit does not exist.

If you try just to plug $x = 3$ into the expression and evaluate it, you'll end up with zero over zero, which is undefined. The trick is first to factor the numerator and denominator and then to cancel factors they have in common:

$$\frac{x^3 + x^2 - 12x}{x^2 - 9} = \frac{x(x^2 + x - 12)}{(x+3)(x-3)}$$

$$= \frac{x(x+4)(x-3)}{(x+3)(x-3)}$$

$$= \frac{x(x+4)}{x+3}$$

Now plug in $x = 3$:

$$\frac{x(x+4)}{x+3} = \frac{(3)(7)}{6} = \frac{21}{6} = \frac{7}{2}$$

The answer is (D).

 Limit of an Algebraic Fraction

To find a limit of an algebraic fraction, first factor the numerator and denominator and cancel any factors they have in common. Then plug in and evaluate.

Sequences Formulas

Arithmetic Sequences:

If d is the difference from one term to the next, a_1 is the first term, a_n is the nth term, and S_n is the sum of the first n terms:

$$a_n = a_1 + (n-1)d$$

$$S_n = n\left(\frac{a_1 + a_n}{2}\right)$$

$$= \frac{n}{2}\left[2a_1 + (n-1)d\right]$$

Geometric Sequences:

If r is the ratio between consecutive terms, a_1 is the first term, a_n is the nth term, and S_n is the sum of the first n terms:

$$a_n = a_1 r^{n-1}$$
$$S_n = \frac{a_1 - a_1 r^n}{1 - r}$$

To find the sum S_∞ of an infinite series when $-1 < r < 1$:

$$S_\infty = \frac{a_1}{1 - r}$$

 Sequences

Our final miscellaneous topic is typified by Example 12. It's a topic that entails a lot of formulas.

★**Example 12**

If the first term in a geometric sequence is 3, and if the third term is 48, what is the 11th term?

(A)   228

(B)   528

(C)   110,592

(D) 3,145,728

(E) 12,582,912

This one's a snap if you know the formula. To find the nth term of a geometric sequence, you need to know the first term (here given as 3), and you need to know the ratio r between consecutive terms. You're not given consecutive terms, but rather the first and the third. The second term would be the first term times r, and the third term would be the second term times r, so:

$$a_1 \times r \times r = a_3$$

$$a_1 r^2 = a_3$$

$$3r^2 = 48$$

$$r^2 = 16$$

$$r = \pm 4$$

Notice that the given information allows for two possible values for r, +4 and –4. The sequence might be:

$$3, 12, 48, 192, \dots$$

Or it might be:

$$3, -12, 48, -192, \dots$$

The only difference in the latter is that the second, fourth, and all other even-numbered terms are negative. You're looking for the eleventh term, which is an odd-numbered term, so it will be the same number in either case. For simplicity's sake, use $r = 4$ in the formula:

$$a_n = a_1 r^{n-1}$$
$$= (3)(4^{10})$$
$$= 3,145,728$$

The answer is (D).

If you don't already at least half-know these formulas, they're probably not worth worrying about. But if you can remember them without too much trouble, one of them might come in handy on Test Day.

There's a lot of material in this chapter, but remember that it's just the tip of the content pyramid. You can use the Miscellaneous Topics Follow-Up Test to see how much you've picked up from this chapter, but keep it in perspective. Whichever test you're taking, it's a higher priority to master the material in the chapters preceding this one.

Miscellaneous Topics Follow-Up Test

Level IC: 7 Questions (9 Minutes)
Level IIC: 12 Questions (15 Minutes)

Directions for Level IC: If you are preparing for the SAT II: Mathematics Level IC Test, solve problems 1–7 and choose the best answer from those given. Fill in the oval corresponding to the best answer choice in the grid to the right of each question. (Answers and explanations begin on page 208.)

Directions for Level IIC: If you are preparing for the SAT II: Mathematics Level IIC Test, solve problems 1–12 and choose the best answer from those given. Fill in the oval corresponding to the best answer choice in the grid to the right of each question. (Answers and explanations begin on page 208.)

DO YOUR FIGURING HERE.

1. The "geocenter" of two positive numbers is defined as the positive square root of their product. If the geocenter of 5 and x is 9, what is the value of x ?

    (A)   6.7

    (B)   7.0

    (C)   11.3

    (D)   13.0

    (E)   16.2

    (A) (B) (C) (D) (E)

DO YOUR FIGURING HERE.

2. "If A is true, then B is false" is logically equivalent to which of the following?
    I. If A is false, then B is true.
    II. If B is false, then A is true.
    III. If B is true, then A is false.

(A) I only

(B) III only

(C) I and II only

(D) II and III only

(E) I, II, and III

3. If $i^2 = -1$, then all of the following are equal EXCEPT

(A) $-i^2$

(B) $(-i)^2$

(C) i^4

(D) $(-i)^4$

(E) $-i^6$

4. To convert a temperature reading from degrees Fahrenheit to degrees Regis, multiply the Fahrenheit reading by $\dfrac{12}{5}$ and subtract 216 from the result. Which of the following represents the Fahrenheit reading that is equivalent to a Regis reading of R degrees?

(A) $\dfrac{5}{12}R + 90$

(B) $\dfrac{5}{12}R - 216$

(C) $\dfrac{5}{12}R + 216$

(D) $\dfrac{12}{5}R - 216$

(E) $\dfrac{12}{5}R + 216$

5. The original price P of a certain item is first discounted by 20 percent, and then 5 percent of the discount price is added for sales tax. If the final price, including the sales tax, is $71.40, what was the original price P?

(A) $59.98

(B) $81.40

(C) $84.00

(D) $85.00

(E) $86.40

Ⓐ Ⓑ Ⓒ Ⓓ Ⓔ

6. The average (arithmetic mean) of all the grades on a certain algebra test was 90. If the average of the 8 males' grades was 87, and the average of the females' grades was 92, how many females took the test?

(A)   8

(B)   9

(C)   10

(D)   11

(E)   12

Ⓐ Ⓑ Ⓒ Ⓓ Ⓔ

7. Rachel drives one-third of the distance from A to B at x kilometers per hour, and she drives the other two-thirds of the distance at y kilometers per hour. What is Rachel's average rate of speed, in kilometers per hour and in terms of x and y, for the entire trip?

(A) $\dfrac{x+y}{2}$

(B) $\dfrac{x+2y}{3}$

(C) $\dfrac{2x+y}{3}$

(D) $\dfrac{xy}{x+y}$

(E) $\dfrac{3xy}{2x+y}$

Ⓐ Ⓑ Ⓒ Ⓓ Ⓔ

DO YOUR FIGURING HERE.

*8. How many ways are there to assign 3 people to 5 desks, with no more than one person to a desk?

(A)   8

(B)   15

(C)   30

(D)   60

(E) 120

*9. A bag contains 10 balls, each labeled with a different integer from 1 to 10, inclusive. If 2 balls are drawn from the bag at random, what is the probability that the sum of the integers on the balls drawn will be greater than 6 ?

(A) 0.41

(B) 0.43

(C) 0.60

(D) 0.76

(E) 0.87

*10. If $\log_x 6 = 3$, then $x =$

(A) 0.500

(B) 1.442

(C) 1.732

(D) 1.817

(E) 2.000

★11. $\lim\limits_{x \to \frac{1}{2}} \dfrac{4x^2 + 8x - 5}{1 - 4x^2} =$

DO YOUR FIGURING HERE.

    (A) -3

    (B) $-\dfrac{1}{2}$

    (C) 0

    (D) $\dfrac{1}{2}$

    (E) 3

Ⓐ Ⓑ Ⓒ Ⓓ Ⓔ

★12. What is the sum of the infinite geometric series

$2 + \left(-\dfrac{1}{2}\right) + \left(\dfrac{1}{8}\right) + \left(-\dfrac{1}{32}\right) + \cdots$?

    (A) $1\dfrac{3}{8}$

    (B) $1\dfrac{2}{5}$

    (C) $1\dfrac{1}{2}$

    (D) $1\dfrac{3}{5}$

    (E) $1\dfrac{5}{8}$

Ⓐ Ⓑ Ⓒ Ⓓ Ⓔ

STOP!   END OF TEST. DO NOT TURN THE PAGE UNTIL
YOU ARE READY TO CHECK YOUR ANSWERS.

Follow-Up Test—Answers and Explanations

Answer Key   1. E 2. B 3. B 4. A 5. D 6. E 7. E 8. D 9. E 10. D 11. A 12. D

1. **(E)**—If 9 is the geocenter of 5 and x, then 9 is equal to the positive square root of the product of 5 and x :

$$9 = \sqrt{5x}$$
$$9^2 = \left(\sqrt{5x}\right)^2$$
$$81 = 5x$$
$$x = \frac{81}{5} = 16.2$$

2. **(B)**—Statement I is not necessarily true: It just negates p and q. (The negation of "B is false," is "B is true.") Statement II is not necessarily true: It just switches p and q. Statement III *is* true: It's the contrapositive.

3. **(B)**—Evaluate each choice:

(A) $-i^2 = (-1)(i^2) = (-1)(-1) = 1$

(B) $(-i)^2 = (-i)(-i) = i^2 = -1$

(C) $i^4 = (i^2)(i^2) = (-1)(-1) = 1$

(D) $(-i)^4 = (-i)^2(-i)^2 = (i^2)(i^2) = (-1)(-1) = 1$

(E) $-i^6 = (-1)(i^6) = (-1)(i^2)(i^2)(i^2) = (-1)(-1)(-1)(-1) = 1$

4. **(A)**—Translate carefully:

$$R = \frac{12}{5}F - 216$$

Then solve for F in terms of R :

$$R + 216 = \frac{12}{5}F$$
$$F = \frac{5}{12}(R + 216)$$
$$F = \frac{5}{12}R + 90$$

5. **(D)**—After a 20% discount, the original price P goes down to $0.8P$. That discount price is then increased by 5%, so the final price is $1.05(0.8P)$. That final price is given as $71.40, so you can set up an equation and solve for the original price P :

$$1.05(0.8P) = 71.40$$
$$0.84P = 71.40$$
$$P = \frac{71.40}{0.84} = 85.00$$

6. **(E)**—The class average is not simply the average of the males' average and the females' average. The class average will be "weighted" in the direction of the larger subgroup, males or females. The class average (90) is closer to the female average (92) than to the male average (87), so there must be more females than males. But how many exactly?

As usual with averages questions, the key is to use the sum. The sum of the 8 males' scores is:

$$\text{Male sum} = (\text{Average})(\text{Number})$$
$$= 87 \times 8$$
$$= 696$$

The sum of the x females' scores is:

$$\text{Female sum} = (\text{Average})(\text{Number})$$

$$= 92x$$

The sum of all the scores is the class average (90) times the total number of people (8 males and x females):

$$\text{Class sum} = (\text{Average})(\text{Number})$$

$$= 90(8 + x)$$

$$= 720 + 90x$$

Now set the sum of the male sum and the female sum equal to the class sum, and solve for x:

$$(\text{Male sum}) + (\text{Female sum}) = \text{Class sum}$$

$$696 + 92x = 720 + 90x$$

$$2x = 24$$

$$x = 12$$

7. (E)—To get the average speed, you need the total distance and the total time. Whatever you call the distance, it will drop out eventually, so you can call it whatever you want—a letter, an expression, a number. You might as well make it something easy to work with. The question stem speaks of "one-third" and "two-thirds" the distance, so it would be easiest to pick a total distance like $3d$ kilometers. (Another good total distance you might have is 300 kilometers.)

For the first d kilometers, Rachel drives at x kilometers per hour. Distance equals rate times time, so time equals distance over rate and the time for the first leg is $\dfrac{d}{x}$ hours. For the next $2d$ kilometers, Rachel drives at y kilometers per hour, so the time for the second leg is $\dfrac{2d}{y}$ hours:

$$\text{Average rate} = \frac{\text{Total distance}}{\text{Total time}}$$

$$= \frac{3d}{\dfrac{d}{x} + \dfrac{2d}{y}}$$

$$= \frac{3d}{\dfrac{dy}{xy} + \dfrac{2dx}{xy}}$$

$$= \frac{3d}{\dfrac{dy + 2dx}{xy}}$$

$$= \frac{3d}{1} \times \frac{xy}{dy + 2dx}$$

$$= \frac{3dxy}{dy + 2dx}$$

$$= \frac{3xy}{2x + y}$$

★8. (D)—This is a permutations question with a twist. Think of it as the arrangement of five items: three distinct people and two indistinguishable non-persons.

$$\text{Number of permutations} = \frac{n!}{a!}$$

$$= \frac{5!}{2!}$$

$$= \frac{5 \times 4 \times 3 \times 2 \times 1}{2 \times 1}$$

$$= 5 \times 4 \times 3$$

$$= 60$$

★9. (E)—To find the probability, you need to know the number of favorable outcomes and the total number of possible outcomes. The latter is the number of combinations of 10 items taken 2 at a time:

$$_{10}C_2 = \frac{10!}{8!2!} = \frac{10 \times 9}{2} = 45$$

To find the number of favorable outcomes, it's easiest to find the number of *un*favorable outcomes (that is, where the numbers add up to 6 or less) by listing and counting them. These are the only unfavorable combinations:

1 and 2	1 and 5
1 and 3	2 and 3
1 and 4	2 and 4

All the other $45 - 6 = 39$ outcomes are favorable, and the probability is $39 \div 45 \approx 0.87$.

*10. **(D)**—Reexpress the equation in exponential form. The base is x, the exponent is 3, and the result is 6:

$$\log_x 6 = 3$$

$$x^3 = 6$$

$$x = \sqrt[3]{6} \approx 1.817$$

*11. **(A)**—First factor the numerator and denominator and cancel common factors:

$$\frac{4x^2 + 8x - 5}{1 - 4x^2} = \frac{(2x - 1)(2x + 5)}{(1 - 2x)(1 + 2x)}$$

$$= \frac{-(1 - 2x)(2x + 5)}{(1 - 2x)(1 + 2x)}$$

$$= \frac{-(2x + 5)}{1 + 2x}$$

$$= \frac{-2x - 5}{2x + 1}$$

Then plug in $x = \frac{1}{2}$:

$$\frac{-2x - 5}{2x + 1} = \frac{-2\left(\frac{1}{2}\right) - 5}{2\left(\frac{1}{2}\right) + 1}$$

$$= \frac{-1 - 5}{1 + 1}$$

$$= \frac{-6}{2} = -3$$

*12. **(D)**—Use the formula. The first term a_1 is 2, and the ratio between consecutive terms is:

$$r = \frac{a_2}{a_1} = \frac{-\frac{1}{2}}{2} = -\frac{1}{4}$$

Plug $a_1 = 2$ and $r = -\frac{1}{4}$ into the formula:

$$\text{Sum} = \frac{a_1}{1 - r}$$

$$= \frac{2}{1 - \left(-\frac{1}{4}\right)} = \frac{2}{\frac{5}{4}}$$

$$= \frac{2}{1} \times \frac{4}{5} = \frac{8}{5} = 1\frac{3}{5}$$

What's Next

Congratulations! You've reviewed the math topics and strategies you need for the SAT II: Mathematics Subject Test. Kaplan's full-length practice tests in the next section will give you a chance to build your test-taking skills even further.

Level IC Practice Tests

This section contains two full-length Kaplan SAT II: Mathematics Level IC Practice Tests. Answers and explanations appear after each test. Scoring information can be found in "Compute Your Level IC Score" at the end of the section.

Kaplan's Level IC
Practice Test A

- The test that follows offers realistic practice for the SAT II: Mathematics Level IC Test. To get the most out of it, you should take it under testlike conditions.

- Take the test in a quiet room with no distractions. Bring some No. 2 pencils and your calculator.

- Time yourself. You should spend no more than one hour on the 50 questions.

- Use the answer sheet to mark your answers.

- Answers and explanations follow the test.

- Scoring instructions are in "Compute your Level IC Score" at the back of this section.

LEVEL IC

KAPLAN PRACTICE TEST A
ANSWER SHEET

1 Ⓐ Ⓑ Ⓒ Ⓓ Ⓔ	14 Ⓐ Ⓑ Ⓒ Ⓓ Ⓔ	27 Ⓐ Ⓑ Ⓒ Ⓓ Ⓔ	40 Ⓐ Ⓑ Ⓒ Ⓓ Ⓔ
2 Ⓐ Ⓑ Ⓒ Ⓓ Ⓔ	15 Ⓐ Ⓑ Ⓒ Ⓓ Ⓔ	28 Ⓐ Ⓑ Ⓒ Ⓓ Ⓔ	41 Ⓐ Ⓑ Ⓒ Ⓓ Ⓔ
3 Ⓐ Ⓑ Ⓒ Ⓓ Ⓔ	16 Ⓐ Ⓑ Ⓒ Ⓓ Ⓔ	29 Ⓐ Ⓑ Ⓒ Ⓓ Ⓔ	42 Ⓐ Ⓑ Ⓒ Ⓓ Ⓔ
4 Ⓐ Ⓑ Ⓒ Ⓓ Ⓔ	17 Ⓐ Ⓑ Ⓒ Ⓓ Ⓔ	30 Ⓐ Ⓑ Ⓒ Ⓓ Ⓔ	43 Ⓐ Ⓑ Ⓒ Ⓓ Ⓔ
5 Ⓐ Ⓑ Ⓒ Ⓓ Ⓔ	18 Ⓐ Ⓑ Ⓒ Ⓓ Ⓔ	31 Ⓐ Ⓑ Ⓒ Ⓓ Ⓔ	44 Ⓐ Ⓑ Ⓒ Ⓓ Ⓔ
6 Ⓐ Ⓑ Ⓒ Ⓓ Ⓔ	19 Ⓐ Ⓑ Ⓒ Ⓓ Ⓔ	32 Ⓐ Ⓑ Ⓒ Ⓓ Ⓔ	45 Ⓐ Ⓑ Ⓒ Ⓓ Ⓔ
7 Ⓐ Ⓑ Ⓒ Ⓓ Ⓔ	20 Ⓐ Ⓑ Ⓒ Ⓓ Ⓔ	33 Ⓐ Ⓑ Ⓒ Ⓓ Ⓔ	46 Ⓐ Ⓑ Ⓒ Ⓓ Ⓔ
8 Ⓐ Ⓑ Ⓒ Ⓓ Ⓔ	21 Ⓐ Ⓑ Ⓒ Ⓓ Ⓔ	34 Ⓐ Ⓑ Ⓒ Ⓓ Ⓔ	47 Ⓐ Ⓑ Ⓒ Ⓓ Ⓔ
9 Ⓐ Ⓑ Ⓒ Ⓓ Ⓔ	22 Ⓐ Ⓑ Ⓒ Ⓓ Ⓔ	35 Ⓐ Ⓑ Ⓒ Ⓓ Ⓔ	48 Ⓐ Ⓑ Ⓒ Ⓓ Ⓔ
10 Ⓐ Ⓑ Ⓒ Ⓓ Ⓔ	23 Ⓐ Ⓑ Ⓒ Ⓓ Ⓔ	36 Ⓐ Ⓑ Ⓒ Ⓓ Ⓔ	49 Ⓐ Ⓑ Ⓒ Ⓓ Ⓔ
11 Ⓐ Ⓑ Ⓒ Ⓓ Ⓔ	24 Ⓐ Ⓑ Ⓒ Ⓓ Ⓔ	37 Ⓐ Ⓑ Ⓒ Ⓓ Ⓔ	50 Ⓐ Ⓑ Ⓒ Ⓓ Ⓔ
12 Ⓐ Ⓑ Ⓒ Ⓓ Ⓔ	25 Ⓐ Ⓑ Ⓒ Ⓓ Ⓔ	38 Ⓐ Ⓑ Ⓒ Ⓓ Ⓔ	
13 Ⓐ Ⓑ Ⓒ Ⓓ Ⓔ	26 Ⓐ Ⓑ Ⓒ Ⓓ Ⓔ	39 Ⓐ Ⓑ Ⓒ Ⓓ Ⓔ	

right

wrong

Use the answer key following the test to count up the number of questions you got right and the number you got wrong. (Remember not to count omitted questions as wrong.) "Compute Your Level IC Score" at the back of this section will show you how to find your score.

LEVEL IC

PRACTICE TEST A

50 Questions (1 hour)

Directions: For each question, choose the BEST answer from the choices given. If the precise answer is not among the choices, choose the one that best approximates the answer. Then fill in the corresponding oval on the answer sheet.

Notes:

(1) To answer some of these questions you will need a calculator. You must use at least a scientific calculator, but programmable and graphing calculators are also allowed.

(2) All angle measures on this test are in degrees, so your calculator should be set to degree mode.

(3) Figures in this test are drawn as accurately as possible UNLESS it is stated in a specific question that the figure is not drawn to scale. All figures are assumed to lie in a plane unless otherwise specified.

(4) The domain of any function f is assumed to be the set of all real numbers x for which $f(x)$ is a real number, unless otherwise indicated.

Reference Information: Use the following formulas as needed.

Right circular cone: If r = radius and h = height, then **Volume** $= \dfrac{1}{3}\pi r^2 h$; and if c = circumference of the base and ℓ = slant height, then **Lateral Area** $= \dfrac{1}{2}c\ell$.

Sphere: If r = radius, then **Volume** $= \dfrac{4}{3}\pi r^3$ and **Surface Area** $= 4\pi r^2$.

Pyramid: If B = area of the base and h = height, then **Volume** $= \dfrac{1}{3}Bh$.

DO YOUR FIGURING HERE.

1.   If $2a + 3 = 6$, then $\dfrac{3}{4a+6} =$

   (A) $\dfrac{1}{4}$ (B) $\dfrac{1}{2}$ (C) 1   (D) 2   (E) 3

TURN TO THE NEXT PAGE.

LEVEL IC

DO YOUR FIGURING HERE.

2. In terms of x, what is the average (arithmetic mean) of $4x - 2$, $x + 2$, $2x + 3$, and $x + 1$?

   (A) $2x - 1$

   (B) $2x$

   (C) $2x + 1$

   (D) $2x + 4$

   (E) $8x + 4$

3. If $4^{2x+2} = 64$, then $x =$

   (A) $\dfrac{1}{2}$ (B) 1   (C) $\dfrac{3}{2}$ (D) 2   (E) $\dfrac{5}{2}$

4. What is the least positive integer that is divisible by both 2 and 5 and leaves a remainder of 2 when it is divided by 7 ?

   (A) 20   (B) 30   (C) 50   (D) 65   (E) 75

5. In Figure 1, the area of rectangle $CDEF$ is twice the area of rectangle $ABCF$. If $CD = 2x + 2$, what is the length of AE, in terms of x ?

   (A) $2x + 3$

   (B) $2x + 4$

   (C) $3x + 1$

   (D) $3x + 2$

   (E) $3x + 3$

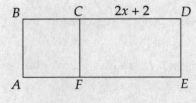

Figure 1

6. If $2y^2 + x - 4 = 0$ and $\dfrac{x}{2} = y^2$, then $x =$

   (A) 1   (B) 2   (C) 3   (D) 4   (E) 5

TURN TO THE NEXT PAGE.

KAPLAN PRACTICE TEST A

DO YOUR FIGURING HERE.

7. If a laser printer can print x pages per minute, how many minutes, in terms of x, would it take the laser printer to print a 100-page document?

    (A) $100x$

    (B) $100 - x$

    (C) $100 + x$

    (D) $\dfrac{x}{100}$

    (E) $\dfrac{100}{x}$

8. In the table, $f(x)$ is a linear function. What is the value of k?

    (A) 3
    (B) 4
    (C) 5
    (D) 6
    (E) 7

x	$f(x)$
0	−4
1	−1
2	2
3	k
4	8

9. Jackie uses 30 percent of her monthly earnings for rent and 50 percent of the remaining amount for food and transportation. If she spends $525 for food and transportation, how much does she pay for rent?

    (A) $400   (B) $450   (C) $500   (D) $550   (E) $600

TURN TO THE NEXT PAGE.

LEVEL IC

KAPLAN PRACTICE TEST A

10. In Figure 2, if congruent right triangles ABD and DCA share leg AD, then $x =$

    (A) 90
    (B) 100
    (C) 110
    (D) 120
    (E) 130

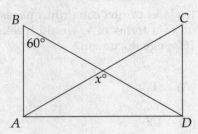

Figure 2

11. If $\dfrac{x+1}{2} + \dfrac{4x-1}{4} = 5.5$, then $x =$

    (A) 2.5
    (B) 3.0
    (C) 3.5
    (D) 4.0
    (E) 4.5

12. If $a \downarrow b = \sqrt[b]{a}$, then $10 \downarrow 3 =$

    (A) 1.12
    (B) 1.69
    (C) 2.15
    (D) 2.71
    (E) 3.33

13. Which of the following ordered pairs is the solution to the equations $2y + x = 5$ and $-2y + x = 9$?

    (A) (1, 7)
    (B) (–1, 7)
    (C) (7, –1)
    (D) (–7, 1)
    (E) (–7, –1)

TURN TO THE NEXT PAGE.

DO YOUR FIGURING HERE.

14. What is the solution set for the equation $|2x - 3| = 13$?

    (A)   {–8}

    (B)   {–5}

    (C)   {–5, –8}

    (D)   {–5, 8}

    (E)   {5, –8}

15. $\dfrac{6!}{2!3!} =$

    (A)   1

    (B)   6

    (C)   15

    (D)   30

    (E)   60

16. In Figure 3, the length of AC is 3 times the length of CD. If B is the midpoint of AC, and the length of CD is 5, what is the length of BD ?

    (A)   10

    (B)   12.5

    (C)   13.5

    (D)   15

    (E)   17.5

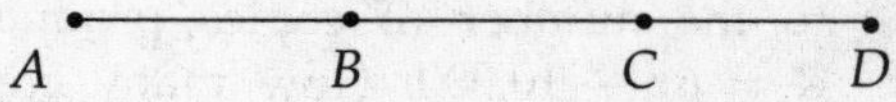

A   B   C   D

Figure 3

17. Which of the following lines is parallel to $y = -2x + 3$ and has a y-intercept of 4 ?

    (A) $y = -2x + 4$

    (B) $y = -2x - 4$

    (C) $y = 2x - 4$

    (D) $y = 2x + 4$

    (E) $y = \dfrac{1}{2}x + 4$

TURN TO THE NEXT PAGE.

DO YOUR FIGURING HERE.

18. In Figure 4, the area of quadrilateral *ABCD* is

    (A)   32
    (B)   33
    (C)   34
    (D)   35
    (E)   36

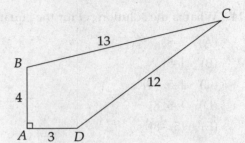

Figure 4

19. If $f(x) = x^2 + x$ and $g(x) = \sqrt{x}$, then $f(g(3)) =$

    (A)   1.73
    (B)   3.46
    (C)   4.73
    (D)   7.34
    (E)   12.00

20. At a certain software company, the cost, *C*, of developing and producing a computer software program is related to the number of copies produced, *x*, by the equation $C = 30,000 + 2x$. The company's total revenues, *R*, is related to the number of copies produced, *x*, by the equation $R = 6x - 10,000$. How many copies must the company produce so that the total revenue is equal to the cost?

    (A)   5,000
    (B)   6,000
    (C)   7,500
    (D)   9,000
    (E)   10,000

TURN TO THE NEXT PAGE.

DO YOUR FIGURING HERE.

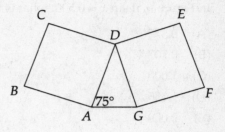

Figure 5

21. If the two squares shown in Figure 5 are identical, what is the degree measure of angle *ADE* ?

    (A) 120
    (B) 135
    (C) 150
    (D) 165
    (E) 175

22. Points $A(\sqrt{2},4)$, $B(6,-\sqrt{3})$, and *C* are collinear. If *B* is the midpoint of line segment *AC*, approximately what are the (x, y) coordinates of point *C* ?

    (A) (3.71, 1.13)
    (B) (3.71, 5.73)
    (C) (7.41, –7.46)
    (D) (10.59, –7.46)
    (E) (10.59, 5.73)

23. What is the solution set to the equation $4 + x^2 = 2x^2 - 5$?

    (A) $\{x: x = 3\}$
    (B) $\{x: x = -3\}$
    (C) $\{x: x = \pm 3\}$
    (D) $\{x: x = -1\}$
    (E) $\{x: x = 1\}$

24. Which of the following triplets can be the lengths of the sides of a triangle?

    (A) 2, 3, 5
    (B) 1, 4, 2
    (C) 7, 4, 4
    (D) 5, 6, 12
    (E) 9, 20, 8

TURN TO THE NEXT PAGE.

DO YOUR FIGURING HERE.

25. In Figure 6, if sin x = 0.500, what is the value of tan x ?

(A)   0.577

(B)   0.707

(C)   1.000

(D)   1.155

(E)   2.000

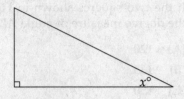

Figure 6

26. In Figure 7, if line ℓ has a slope of 1 and passes through the origin, which of the following points has (x, y) coordinates such that $\dfrac{x}{y} > 1$?

(A) A

(B) B

(C) C

(D) D

(E) E

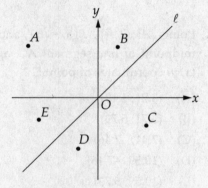

Figure 7

27. On a recent chemistry test, the average (arithmetic mean) score among 5 students was 83, where the lowest and highest possible scores are 0 and 100, respectively. If the teacher decides to increase each student's score by 2 points, and if none of the students originally scored more than 98, which of the following must be true?

  I. After the scores are increased, the average score is 85.

 II. When the scores are increased, the difference between the highest and lowest scores increases.

 III. After the increase, all 5 scores are greater than or equal to 25.

(A)   I only

(B)   II only

(C)   I and II only

(D)   I and III only

(E)   I, II, and III

TURN TO THE NEXT PAGE.

LEVEL IC

DO YOUR FIGURING HERE.

28. If $a > b$ and $c > d$, which of the following must be true?

    (A) $ac > bd$

    (B) $a + b > c + d$

    (C) $a + c > b + d$

    (D) $a - b > c - d$

    (E) $ad > bc$

29. $1 - 2\sin^2\theta - 2\cos^2\theta =$

    (A) -2

    (B) -1

    (C) 0

    (D) 1

    (E) 2

30. Sheila leaves her house and starts driving due south for 30 miles, then drives due west for 60 miles, and finally drives due north for 10 miles to reach her office. Which of the following is the approximate straight-line distance, in miles, from her house to her office?

    (A) 63   (B) 67   (C) 71   (D) 75   (E) 80

31. If $f(x) = x^2 - 1$, $g(x) = (x - 1)^{-1}$, and $x \neq 1$, then $f(x)g(x) =$

    (A) $2x + 1$

    (B) $x + 1$

    (C) $x - 1$

    (D) $x^3 - 1$

    (E) $2x - 1$

TURN TO THE NEXT PAGE.

DO YOUR FIGURING HERE.

32. If an empty rectangular water tank that has dimensions 100 centimeters, 20 centimeters, and 40 centimeters is to be filled using a right cylindrical bucket with a base radius of 9 centimeters and a height of 20 centimeters, approximately how many buckets of water will it take to fill the tank?

    (A) 14   (B) 16   (C) 18   (D) 20   (E) 22

33. Sarah is scheduling the first four periods of her school day. She needs to fill those periods with calculus, art, literature, and physics, and each of these courses is offered during each of the first four periods. How many different schedules can Sarah choose from?

    (A) 1   (B) 4   (C) 12   (D) 24   (E) 120

34. In Figure 8, AE is parallel to BD. What is the length of DE ?

    (A)   2.33
    (B)   2.67
    (C)   3.33
    (D)   3.67
    (E)   6.67

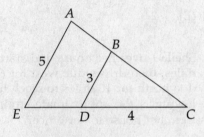

Figure 8

35. If $f(x) = \sqrt{x^2 - 4}$, what is the domain?

    (A)   All real numbers
    (B)   All x such that $x \geq 2$
    (C)   All x such that $x \leq -2$
    (D)   All x such that $-2 \leq x \leq 2$
    (E)   All x such that $x \leq -2$ or $x \geq 2$

36. What is the area of a triangle with vertices $(1, 1)$, $(3, 1)$, and $(5, 7)$?

    (A) 6   (B) 7   (C) 9   (D) 10   (E) 12

TURN TO THE NEXT PAGE.

KAPLAN PRACTICE TEST A

DO YOUR FIGURING HERE.

37. Which of the following inequalities is equivalent to $-2(x + 5) < -4$?

    (A) $x > -3$

    (B) $x < -3$

    (C) $x > 3$

    (D) $x < 3$

    (E) $x > 7$

·38. If $i = \sqrt{-1}$, for which of the following values of n does $i^{n} + (-i)^{n}$ have a positive value?

    (A) 23   (B) 24   (C) 25   (D) 26   (E) 27

39. The maximum value of the function $f(x) = 1 - \cos x$ between 0 and 2π is

    (A) 0   (B) 1   (C) 1.5   (D) 2   (E) 2.5

40. In Figure 9, if $y > 60$ and $AB = BC$, which of the following must be true?

    I. $a + y = 180$

    II. $y > z$

    III. $a = y + z$

    (A) I only

    (B) II only

    (C) III only

    (D) I and II

    (E) II and III

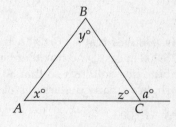

Figure 9

TURN TO THE NEXT PAGE.

DO YOUR FIGURING HERE.

41. If $f(x) = \dfrac{1}{x}$, and $0 < x < 1$, what is the range of $f(x)$?

    (A)   All real numbers

    (B)   All real numbers between 0 and 1

    (C)   All real numbers greater than 0

    (D)   All real numbers greater than 1

    (E)   All real numbers greater than or equal to 1

42. Two identical spheres of radius 6 intersect so that the distance between their centers is 10. The points of intersection of the two spheres form a circle. What is the area of this circle?

    (A) 5π (B) 6π (C) 8π (D) 10π (E) 11π

43. If a remainder of 4 is obtained when $x^3 + 2x^2 - x - k$ is divided by $x - 2$, what is the value of k?

    (A) 4   (B) 6   (C) 10   (D) 12   (E) 14

44. Ms. Hobbes has a portfolio that includes $50,000 in stock, $75,000 in cash, and no other holdings. If she wishes to redistribute her holdings so that 80 percent of the portfolio is in cash, how many dollars of stock must she convert to cash?

    (A)   10,000

    (B)   15,000

    (C)   20,000

    (D)   25,000

    (E)   30,000

45. A sphere of radius 5 has the same volume as a cube with an edge of what length?

    (A)   5.00

    (B)   5.50

    (C)   6.24

    (D)   8.06

    (E)   9.27

TURN TO THE NEXT PAGE.

LEVEL IC

DO YOUR FIGURING HERE.

46. The equation $x^2 = y^2$ is represented by which of the following graphs?

(A)

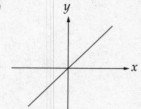

(B)

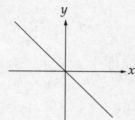

(C)

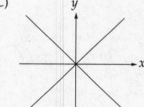

(D)

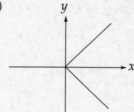

(E)

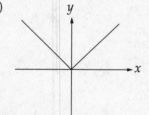

TURN TO THE NEXT PAGE.

KAPLAN PRACTICE TEST A

DO YOUR FIGURING HERE.

47. If point $A(3, 5)$ is located on a circle in the coordinate plane, and the center of the circle is the origin, which of the following points must lie outside this circle?

    (A) $(1.0, 6.0)$
    (B) $(1.5, 5.5)$
    (C) $(2.5, 4.5)$
    (D) $(4.0, 4.0)$
    (E) $(5.0, 3.0)$

48. If $f(x) = 3x - 1$, n represents the slope of the line with the equation $y = f^{-1}(x)$, and p represents the slope of a line that is perpendicular to the line with the equation $y = f(x)$, then $np =$

    (A) -9

    (B) $-\dfrac{1}{9}$

    (C) $\dfrac{1}{9}$

    (D) 9

    (E) It cannot be determined from the information given.

TURN TO THE NEXT PAGE.

49. The parabola with the equation $y = 4x - \dfrac{1}{2}x^2$ has how many points with (x, y) coordinates that are both positive integers?

    (A) 3

    (B) 4

    (C) 7

    (D) 8

    (E) Infinitely many

DO YOUR FIGURING HERE.

50. In Figure 10, each of the 3 circles is tangent to the other 2, and each side of the equilateral triangle is tangent to 2 of the circles. If the length of one side of the triangle is x, what is the radius, in terms of x, of one of the circles?

    (A) $\dfrac{x}{1 + 2\sqrt{3}}$

    (B) $\dfrac{x}{2 + 2\sqrt{3}}$

    (C) $\dfrac{x}{1 + \sqrt{3}}$

    (D) $\dfrac{2x}{1 + \sqrt{3}}$

    (E) $\dfrac{2x}{1 + 2\sqrt{3}}$

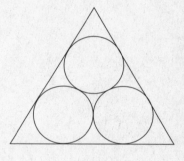

Figure 10

STOP! END OF TEST. DO NOT TURN THE PAGE UNTIL YOU ARE READY TO CHECK YOUR ANSWERS.

Level IC Test A—Answer Key

1. A	11. C	21. A	31. B	41. D
2. C	12. C	22. D	32. B	42. E
3. A	13. C	23. C	33. D	43. C
4. B	14. D	24. C	34. B	44. D
5. E	15. E	25. A	35. E	45. D
6. B	16. B	26. E	36. A	46. C
7. E	17. A	27. D	37. A	47. A
8. C	18. E	28. C	38. B	48. B
9. B	19. C	29. B	39. D	49. A
10. D	20. E	30. A	40. E	50. B

1. **(A)**—You don't need to find the value of a to answer this question. It's more direct to recognize that the denominator of the expression you're solving for is exactly twice the left side of the given equation:

$$2a + 3 = 6$$
$$2(2a + 3) = 2(6)$$
$$4a + 6 = 12$$
$$\frac{1}{4a + 6} = \frac{1}{12}$$
$$\frac{3}{4a + 6} = \frac{3}{12} = \frac{1}{4}$$

2. **(C)**—To find the average of four expressions, add them up and divide by 4. The sum is $8x + 4$. When you divide that by 4, you get $2x + 1$.

3. **(A)**—To solve an equation with the unknown in an exponent, reexpress the equation so that both sides have the same base:

$$4^{2x+2} = 64$$
$$4^{2x+2} = 4^3$$
$$2x + 2 = 3$$
$$2x = 1$$
$$x = \frac{1}{2}$$

4. **(B)**—The easiest way to find the answer to this question is to check the answer choices, starting with the smallest, until you find one that meets the given conditions. A multiple of 2 and 5 is a multiple of 10, so already you know that the answer is (A), (B), or (C). Starting with (A), divide each by 7 until you find one that leaves a remainder of 2. (A) 20 divided by 7 is 2 with a remainder of 6—that's not it. (B) 30 divided by 7 is 4 with a remainder of 2—that's it.

5. **(E)**—The two rectangles have the same height, so if $CDEF$ has twice the area of $ABCF$, then the base of $CDEF$ is twice the base of $ABCF$. The longer base is given as $2x + 2$, so the shorter base is half that, or $x + 1$. You're looking for AE, which is the sum of the bases:

$$AE = AF + FE$$
$$= (x + 1) + (2x + 2)$$
$$= 3x + 3$$

6. (B)—The second equation expresses y^2 in terms of x, so you can substitute that expression for y^2 in the first equation and solve for x:

$$2y^2 + x - 4 = 0$$
$$2\left(\frac{x}{2}\right) + x - 4 = 0$$
$$x + x - 4 = 0$$
$$2x = 4$$
$$x = 2$$

7. (E)— A rate of x pages per minute is the same as 100 pages per how many minutes? If y is the number of minutes it takes to print 100 pages, then you can set up the proportion:

$$\frac{x \text{ pages}}{1 \text{ minute}} = \frac{100 \text{ pages}}{y \text{ minutes}}$$
$$xy = 100$$
$$y = \frac{100}{x}$$

8. (C)—The given values of x are evenly spaced, and $f(x)$ is a linear (i.e., straight-line) function, so the values of $f(x)$ will be evenly spaced. The unknown k is halfway between 2 and 8, so $k = 5$.

9. (B)—After spending 30% on rent, Jackie has 70% of her earnings left. Half of that 70% is 35%, which goes for food and transportation. So the ratio of rent to food and transportation is 30:35, or 6:7. Now you can set up a proportion and solve for the rent:

$$\frac{\text{rent}}{\text{food \& trans}} = \frac{6}{7}$$
$$\frac{\text{rent}}{\$525} = \frac{6}{7}$$
$$\text{rent} = 6 \times \$525 \div 7 = \$450$$

10. (D)—Mark up the figure. Because the triangles are congruent, $\angle C$ also measures 60°. And because they're both right triangles, $\angle BDA$ and $\angle CAD$ both measure 30°:

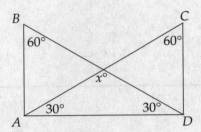

Now you have the other two angles inside a triangle with x, so you can set their sum equal to 180 and solve for x:

$$x + 30 + 30 = 180$$
$$x = 120$$

11. (C)—To solve an equation with fractions, first multiply both sides by whatever you have to to clear all denominators. In this case, multiply both sides by 4:

$$\frac{x+1}{2} + \frac{4x-1}{4} = 5.5$$
$$4\left(\frac{x+1}{2} + \frac{4x-1}{4}\right) = 4(5.5)$$
$$(2x+2) + (4x-1) = 22$$
$$6x + 1 = 22$$
$$6x = 21$$
$$x = \frac{21}{6} = 3.5$$

12. (C)—Plug $a = 10$ and $b = 3$ into the definition:

$$a \downarrow b = \sqrt[b]{a}$$
$$10 \downarrow 3 = \sqrt[3]{10} \approx 2.15$$

13. (C)—If you add the equations as presented, the y's drop out:

$$2y + x = 5$$
$$\underline{-2y + x = 9}$$
$$2x = 14$$
$$x = 7$$

Then plug $x = 7$ back into one of the original equations to find y:

$$2y + x = 5$$
$$2y + 7 = 5$$
$$2y = -2$$
$$y = -1$$

14. (D)—To solve an equation with absolute value, consider the two possibilities. If $|2x - 3| = 13$, then what's inside the absolute value signs equals either 13 or –13:

$$2x - 3 = 13$$
$$2x = 16$$
$$x = 8$$

OR:

$$2x - 3 = -13$$
$$x = -10$$
$$x = -5$$

15. (E)—Expand the factorials, cancel factors common to the numerator and denominator, and then calculate:

$$\frac{6!}{2!3!} = \frac{6 \times 5 \times 4 \times 3 \times 2 \times 1}{2 \times 1 \times 3 \times 2 \times 1}$$
$$= \frac{6 \times 5 \times 4}{2} = 60$$

16. (B)—Mark up the figure. $CD = 5$ and AC is 3 times that, so $AC = 15$:

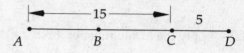

B is the midpoint of AC, so $AB = BC = 7.5$:

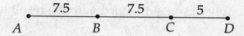

Now you can see that $BD = BC + CD = 7.5 + 5 = 12.5$.

17. (A)—Parallel lines have the same slope, so you're looking for a line with a slope of –2. Only (A) and (B) have that slope, and of those only (A) has y-intercept +4.

18. (E)—Add to the figure. Diagonal BD will divide the quadrilateral into two familiar triangles:

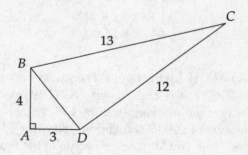

ABD is a right triangle with legs 3 and 4, so $BD = 5$, and therefore BCD is a 5-12-13 right triangle. The area of the 3-4-5 triangle is $\frac{1}{2}(\text{leg}_1)(\text{leg}_2) = \frac{1}{2}(3)(4) = 6$, and the area of the 5-12-13 right triangle is $\frac{1}{2}(\text{leg}_1)(\text{leg}_2) = \frac{1}{2}(5)(12) = 30$. The quadrilateral's area, then, is $6 + 30 = 36$.

19. (C)—Apply the inside function first:

$$g(x) = \sqrt{x}$$
$$g(3) = \sqrt{3}$$

Then apply the outside function to the result:

$$f(x) = x^2 + x$$
$$f(\sqrt{3}) = (\sqrt{3})^2 + \sqrt{3}$$
$$= 3 + \sqrt{3} \approx 4.73$$

20. (E)—Put much more simply, what the question is asking is: At what point does cost equal revenue? In algebraic terms, for what value of x does C equal R ?

$$C = R$$
$$30,000 + 2x = 6x - 10,000$$
$$-4x = -40,000$$
$$x = 10,000$$

21. (A)—Mark up the figure. The squares are identical, so $AD = DG$ and triangle ADG is isosceles. That makes the measure of $\angle DGA$ 75 degrees, and the measure of $\angle ADG$ is therefore $180 - 75 - 75 = 30$ degrees. Now you know every angle in the figure:

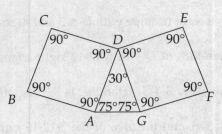

The measure of $\angle ADE$ is $30 + 90 = 120$ degrees.

22. (D)—If $B\left(6, -\sqrt{3}\right)$ is the midpoint of $A\left(\sqrt{2}, 4\right)$ and $C(x, y)$, then 6 is the average of $\sqrt{2}$ and x, and $-\sqrt{3}$ is the average of 4 and y:

$$\frac{\sqrt{2} + x}{2} = 6$$
$$\sqrt{2} + x = 12$$
$$x = 12 - \sqrt{2} \approx 10.59$$

AND:

$$\frac{4 + y}{2} = -\sqrt{3}$$
$$4 + y = -2\sqrt{3}$$
$$y = -2\sqrt{3} - 4 \approx -7.46$$

23. (C)—Simplify the equation:

$$4 + x^2 = 2x^2 - 5$$
$$4 + 5 = 2x^2 - x^2$$
$$9 = x^2$$

Don't jump to the conclusion that if 9 equals x^2, then x must be 3. x could just as well be –3:

$$x^2 = 9$$
$$x = \pm 3$$

24. (C)—To satisfy the Triangle Inequality Theorem, adding the two shorter sides must give you more than the longest side. In (B), (D), and (E), the two shorter sides add up to less than the longest side, and in (A) they add up to the same as the longest side. Only in (C) do they add up to more.

25. (A)—If sin x is .5, then the side opposite x is one-half the hypotenuse. In other words, this is a 30-60-90 triangle with familiar side ratios:

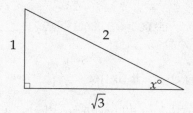

Tangent is opposite over adjacent, so:

$$\tan x = \frac{1}{\sqrt{3}} \approx 0.577$$

26. (E)—When x and y have different signs, that is, when the point is in the second or fourth quadrant, $\frac{x}{y}$ is negative and so for no point in those quadrants can $\frac{x}{y}$ be greater than 1. That eliminates (A) and (C). When x and y are both positive, $\frac{x}{y}$ will be greater than 1 if x is greater than y. Point B is above the line $x = y$, so for that point $y > x$. When x and y are both negative, $\frac{x}{y}$ will be greater than 1 if x is less than y. Point D is below the line $x = y$, so for that point $x > y$. Point E, however, is above the line $x = y$, so for that point $y > x$, and $\frac{x}{y} > 1$.

27. (D)—If the average of the 5 original scores was 83, then the sum of those 5 scores was $5 \times 83 = 415$. When each score is increased by 2, the sum goes up to 425, one-fifth of which is 85. So the new average is 85 and statement I is true. Statement II, however, is *not* true because if both the lowest and highest scores go up by the same amount, the difference between those scores remains the same. So I is true and II is not, and the answer is either (A) or (D). What about III? To find the lowest score that one of the 5 students can get, imagine that the other 4 students all get perfect 100s. Four 100s add up to 400, which leaves $425 - 400 = 25$ for the fifth and lowest possible score for the 5 students. III is true.

28. (C)—You could do this one by picking numbers, but for any set of numbers you might pick, more than one of the answer choices will be true. You will therefore have to pick at least two sets of numbers to find the correct answer. The statement that *must* be true is the one that holds for all a, b, c, and d that fit the given conditions. (A) is tempting: You might think that the product of the larger a and c will be greater than the product of the smaller b and d, and it is if all you pick is positive integers. (A) doesn't hold when you consider negatives. But (C) holds for all possible values of $a, b, c,$ and d. The sum of two larger quantities is greater than the sum of two smaller quantities.

29. (B)—Look for "$\sin^2 + \cos^2$" in the expression:

$$1 - 2\sin^2\theta - 2\cos^2\theta = 1 - 2(\sin^2\theta + \cos^2\theta)$$

$$= 1 - 2 = -1$$

30. **(A)**—Sketch a diagram:

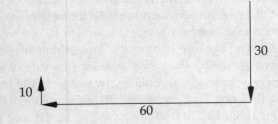

The net change is south 20 and west 60:

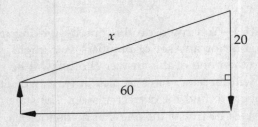

So you're looking for the hypotenuse of a right triangle with legs of 20 and 60:

$$x = \sqrt{20^2 + 60^2}$$
$$= \sqrt{400 + 3,600} = \sqrt{4,000} \approx 63$$

31. **(B)**—This time there's no inside and outside functions. What you're looking for here is the product of the functions:

$$f(x) = x^2 - 1$$
$$g(x) = (x-1)^{-1} = \frac{1}{x-1}$$
$$f(x)g(x) = (x^2 - 1)\left(\frac{1}{x-1}\right)$$
$$= \frac{x^2 - 1}{x-1}$$
$$= \frac{(x-1)(x+1)}{x-1}$$
$$= x + 1$$

32. **(B)**—The volume of the tank is $100 \times 20 \times 40 = 80,000$ cubic centimeters. The volume of the bucket is $\pi r^2 h \approx (3.14)(9^2)(20) = 5,086.8$. To figure out how many buckets fit into the tank, divide:

$$80,000 \div 5,086.8 \approx 15.7$$

The nearest choice is (B) 16.

33. **(D)**— She can fill her first period slot with any of the 4 subjects, leaving 3 for the second, 2 for the third, and 1 for the fourth:

$$4 \times 3 \times 2 \times 1 = 24$$

34. **(B)**—That AE and BD are parallel tells you that the outside triangle ACE and the inside triangle BCD are similar. Corresponding sides are proportional:

$$\frac{BD}{AE} = \frac{CD}{CE}$$
$$\frac{3}{5} = \frac{4}{4+x}$$
$$3(4+x) = 5 \times 4$$
$$12 + 3x = 20$$
$$x = 8$$
$$x = \frac{8}{3} \approx 2.67$$

35. **(E)**—For what values of x can you take the square root of $x^2 - 4$? Only those values for which $x^2 - 4$ is nonnegative:

$$x^2 - 4 \geq 0$$
$$x^2 \geq 4$$
$$x \leq -2 \text{ or } x \geq 2$$

36. (A)—Sketch the diagram:

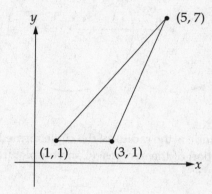

If you use the short side from (1, 1) to (3, 1) as the base, then the base is 2 and the height is 6:

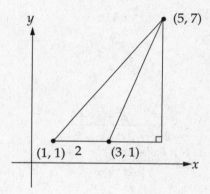

Plug $b = 2$ and $h = 6$ into the formula for the area of a triangle:

$$\text{Area of Triangle} = \frac{1}{2}bh = \frac{1}{2}(2)(6) = 6$$

37. (A)—Simplify:

$$-2(x + 5) < -4$$
$$-2x - 10 < -4$$
$$-2x < -4 + 10$$
$$-2x < 6$$
$$x > \frac{6}{-2}$$
$$x > -3$$

Notice that the inequality sign flipped when both sides were divided by –2.

38. (B)—When you raise i to successive integer exponents, a pattern develops:

$$i^1 = i$$
$$i^2 = -1$$
$$i^3 = -i$$
$$i^4 = 1$$
$$i^5 = i$$

Every fourth power the pattern repeats. A similar pattern develops with $-i$:

$$(-i)^1 = -i$$
$$(-i)^2 = -1$$
$$(-i)^3 = i$$
$$(-i)^4 = 1$$
$$(-i)^5 = -i$$

Whenever the exponent is a multiple of 4, i^n and $(-i)^n$ are both equal to 1. (B) is the only multiple of 4 among the choices. When $n = 24$, $i^n + (-i)^n = 1 + 1 = 2$, which is a positive integer.

39. (D)—Between 0 and 2π, $\cos x$ ranges from –1 to 1. The maximum value of $f(x) = 1 - \cos x$ will be when $\cos x$ is at its least, that is, when $\cos x = -1$:

$$f(x) = 1 - \cos x$$
$$= 1 - (-1) = 1 + 1 = 2$$

40. (E)—The triangle is isosceles; since $AB = BC$, $x = z$. Since $y > 60$, that leaves less than 120 degrees to split between x and z, so x and z are both less than 60. Angle a is an exterior angle, so it's equal to the sum of the remote interior angles:

$$a = x + y$$

Now you can see that II is true, because $y > 60$ and $z < 60$, and you can see that III is true, because $a = x + y$ and $x = z$. I is not true: y is greater than z and $a + z = 180$, so $a + y$ is greater than 180.

41. (D)—Imagine (or use your graphing calculator) the graph of $f(x) = \dfrac{1}{x}$ between 0 and 1. At $x = 1$, $f(x) = 1$. As x gets smaller, $f(x)$ gets bigger. $f\left(\dfrac{1}{2}\right) = 2 \cdot f\left(\dfrac{1}{100}\right) = 100$. As x approaches 0, $f(x)$ gets infinitely large. The range is all real numbers greater than 1.

42. (E)—This one's hard to sketch because it's 3-D. Here's a cross section:

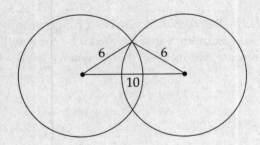

Drop an altitude, and you make two right triangles:

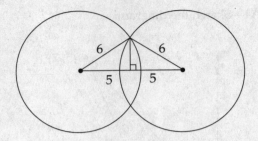

That altitude is the radius of the circle formed by the cross section of the spheres. (Can you picture it?)

$$r = \sqrt{6^2 - 5^2}$$
$$= \sqrt{36 - 25}$$
$$= \sqrt{11}$$
$$\pi r^2 = 11\pi$$

43. (C)

$$
\require{enclose}
\begin{array}{r}
x^2 + 4x + 7 \\
x - 2 \enclose{longdiv}{x^3 + 2x^2 - x - k} \\
\underline{x^3 - 2x^2} \\
4x^2 - x \\
\underline{4x^2 - 8x} \\
7x - k \\
\underline{7x - 14} \\
-k + 14
\end{array}
$$

In algebraic terms, the remainder is $-k + 14$. If that's equal to 4, then:

$$-k + 14 = 4$$
$$10 = k$$

44. (D)—The total portfolio is 50,000 + 75,000 = 125,000 dollars. Eighty percent of that is (.80)(125,000) = 100,000 dollars. Thus she needs to convert 100,000 – 75,000 = 25,000 dollars.

45. (D)—A sphere of radius 5 has volume $\frac{4}{3}\pi(5)^3 \approx 523.6$. If that's also the volume of a cube, then an edge of that cube is the cube root of that. Use your calculator: $\sqrt[3]{523.6} \approx 8.06$.

46. (C)—The equation $x^2 = y^2$ is true whenever $x = \pm y$. Choice (A) shows $x = y$. Choice (B) shows $x = -y$. Choice (C) shows both.

47. (A)—OA is a radius of the circle, and $OA = \sqrt{3^2 + 5^2} = \sqrt{34}$. You're looking for the choice for which $x^2 + y^2$ is greater than 34. Use your calculator. In (A) $x^2 + y^2 = 37$. So it looks like the answer is (A). If you have lots of time, you can check the other answer choices, just to be sure. In (B) $x^2 + y^2 = 32.5$. In (C) $x^2 + y^2 = 26.5$. In (D) $x^2 + y^2 = 32$. And in (E) $x^2 + y^2 = 34$. Only (A) is greater than 34.

48. (B)—You don't need to actually find the inverse function to answer this question if you remember that the slopes of inverse functions are reciprocals. The slope of the given $f(x) = 3x - 1$ is 3, so the slope of $f^{-1}(x)$ is $\frac{1}{3}$. And the slope of a line perpendicular to $f(x)$ will have a slope that's the negative reciprocal of 3, or $-\frac{1}{3}$. So, $n = \frac{1}{3}$ and $p = -\frac{1}{3}$, and therefore $np = \left(\frac{1}{3}\right)\left(-\frac{1}{3}\right) = -\frac{1}{9}$.

49. (A)—The question is asking, in other words, for how many positive integer values of x does y turn out to be a positive integer? The expression $4x - \frac{1}{2}x^2$ will be positive whenever $4x$ is greater than $\frac{1}{2}x^2$:

$$4x > \frac{1}{2}x^2$$
$$\frac{1}{2}x^2 - 4x < 0$$
$$x\left(\frac{1}{2}x - 4\right) < 0$$
$$x > 0 \text{ and } \frac{1}{2}x - 4 < 0$$
$$x > 0 \text{ and } x < 8$$
$$0 < x < 8$$

The expression $4x - \frac{1}{2}x^2$ will be an integer whenever $\frac{1}{2}x^2$ is an integer. Half of x^2 is an integer whenever x is even, so y will be a positive integer if x is an even number greater than 0 and less than 8— in other words, if x is 2, 4, or 6. That's 3 points.

50. (B)—Mark up the figure:

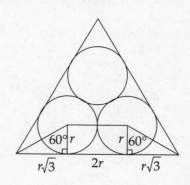

$$x = r\sqrt{3} + 2r + r\sqrt{3}$$
$$x = 2r + 2r\sqrt{3}$$
$$x = r\left(2 + 2\sqrt{3}\right)$$
$$r = \frac{x}{2 + 2\sqrt{3}}$$

Kaplan's Level IC Practice Test B

- The test that follows offers realistic practice for the SAT II: Mathematics Level IC Test. To get the most out of it, you should take it under testlike conditions.

- Take the test in a quiet room with no distractions. Bring some No. 2 pencils and your calculator.

- Time yourself. You should spend no more than one hour on the 50 questions.

- Use the answer sheet to mark your answers.

- Answers and explanations follow the test.

- Scoring instructions are in "Compute your Level IC Score" at the back of this section.

LEVEL IC

1 Ⓐ Ⓑ Ⓒ Ⓓ Ⓔ	14 Ⓐ Ⓑ Ⓒ Ⓓ Ⓔ	27 Ⓐ Ⓑ Ⓒ Ⓓ Ⓔ	40 Ⓐ Ⓑ Ⓒ Ⓓ Ⓔ
2 Ⓐ Ⓑ Ⓒ Ⓓ Ⓔ	15 Ⓐ Ⓑ Ⓒ Ⓓ Ⓔ	28 Ⓐ Ⓑ Ⓒ Ⓓ Ⓔ	41 Ⓐ Ⓑ Ⓒ Ⓓ Ⓔ
3 Ⓐ Ⓑ Ⓒ Ⓓ Ⓔ	16 Ⓐ Ⓑ Ⓒ Ⓓ Ⓔ	29 Ⓐ Ⓑ Ⓒ Ⓓ Ⓔ	42 Ⓐ Ⓑ Ⓒ Ⓓ Ⓔ
4 Ⓐ Ⓑ Ⓒ Ⓓ Ⓔ	17 Ⓐ Ⓑ Ⓒ Ⓓ Ⓔ	30 Ⓐ Ⓑ Ⓒ Ⓓ Ⓔ	43 Ⓐ Ⓑ Ⓒ Ⓓ Ⓔ
5 Ⓐ Ⓑ Ⓒ Ⓓ Ⓔ	18 Ⓐ Ⓑ Ⓒ Ⓓ Ⓔ	31 Ⓐ Ⓑ Ⓒ Ⓓ Ⓔ	44 Ⓐ Ⓑ Ⓒ Ⓓ Ⓔ
6 Ⓐ Ⓑ Ⓒ Ⓓ Ⓔ	19 Ⓐ Ⓑ Ⓒ Ⓓ Ⓔ	32 Ⓐ Ⓑ Ⓒ Ⓓ Ⓔ	45 Ⓐ Ⓑ Ⓒ Ⓓ Ⓔ
7 Ⓐ Ⓑ Ⓒ Ⓓ Ⓔ	20 Ⓐ Ⓑ Ⓒ Ⓓ Ⓔ	33 Ⓐ Ⓑ Ⓒ Ⓓ Ⓔ	46 Ⓐ Ⓑ Ⓒ Ⓓ Ⓔ
8 Ⓐ Ⓑ Ⓒ Ⓓ Ⓔ	21 Ⓐ Ⓑ Ⓒ Ⓓ Ⓔ	34 Ⓐ Ⓑ Ⓒ Ⓓ Ⓔ	47 Ⓐ Ⓑ Ⓒ Ⓓ Ⓔ
9 Ⓐ Ⓑ Ⓒ Ⓓ Ⓔ	22 Ⓐ Ⓑ Ⓒ Ⓓ Ⓔ	35 Ⓐ Ⓑ Ⓒ Ⓓ Ⓔ	48 Ⓐ Ⓑ Ⓒ Ⓓ Ⓔ
10 Ⓐ Ⓑ Ⓒ Ⓓ Ⓔ	23 Ⓐ Ⓑ Ⓒ Ⓓ Ⓔ	36 Ⓐ Ⓑ Ⓒ Ⓓ Ⓔ	49 Ⓐ Ⓑ Ⓒ Ⓓ Ⓔ
11 Ⓐ Ⓑ Ⓒ Ⓓ Ⓔ	24 Ⓐ Ⓑ Ⓒ Ⓓ Ⓔ	37 Ⓐ Ⓑ Ⓒ Ⓓ Ⓔ	50 Ⓐ Ⓑ Ⓒ Ⓓ Ⓔ
12 Ⓐ Ⓑ Ⓒ Ⓓ Ⓔ	25 Ⓐ Ⓑ Ⓒ Ⓓ Ⓔ	38 Ⓐ Ⓑ Ⓒ Ⓓ Ⓔ	
13 Ⓐ Ⓑ Ⓒ Ⓓ Ⓔ	26 Ⓐ Ⓑ Ⓒ Ⓓ Ⓔ	39 Ⓐ Ⓑ Ⓒ Ⓓ Ⓔ	

right

wrong

Use the answer key following the test to count up the number of questions you got right and the number you got wrong. (Remember not to count omitted questions as wrong.) "Compute Your Level IC Score" at the back of this section will show you how to find your score.

LEVEL IC

PRACTICE TEST B

50 Questions (1 hour)

Directions: For each question, choose the BEST answer from the choices given. If the precise answer is not among the choices, choose the one that best approximates the answer.

Notes:

(1) To answer some of these questions you will need a calculator. You must use at least a scientific calculator, but programmable and graphing calculators are also allowed.

(2) All angle measures on this test are in degrees, so your calculator should be set to degree mode.

(3) Figures in this test are drawn as accurately as possible UNLESS it is stated in a specific question that the figure is not drawn to scale. All figures are assumed to lie in a plane unless otherwise specified.

(4) The domain of any function f is assumed to be the set of all real numbers x for which $f(x)$ is a real number, unless otherwise indicated.

Reference Information: Use the following formulas as needed.

Right circular cone: If r = radius and h = height, then **Volume** $= \dfrac{1}{3}\pi r^2 h$; and if c = circumference of the base and ℓ = slant height, then **Lateral Area** $= \dfrac{1}{2}c\ell$.

Sphere: If r = radius, then **Volume** $= \dfrac{4}{3}\pi r^3$ and **Surface Area** $= 4\pi r^2$.

Pyramid: If B = area of the base and h = height, then **Volume** $= \dfrac{1}{3}Bh$.

DO YOUR FIGURING HERE.

1. If $a + b = 10$ and both a and b are positive integers, what is the largest possible value for a ?

   (A) 6 (B) 7 (C) 8 (D) 9 (E) 10

TURN TO THE NEXT PAGE.

KAPLAN PRACTICE TEST B

DO YOUR FIGURING HERE.

2. If $2x + y = 2y - x$, then $y =$

(A) $3x$

(B) $4x$

(C) $2x + 2$

(D) $3x + 1$

(E) $4x + 1$

3. If a certain car can travel 240 miles on 12 gallons of gasoline, then at the same rate, how many gallons of gasoline are needed to travel 300 miles?

(A) 10　　(B) 15　　(C) 20　　(D) 25　　(E) 30

4. If $2(1 + 2x) - 5(4 - 2x) = 14$, $x =$

(A) –5.33　(B) –0.29　(C) 0.29　(D) 0.67　(E) 2.29

5. In Figure 1, if ABC is a straight line, $x =$

(A) 50

(B) 60

(C) 70

(D) 80

(E) 90

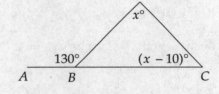

Figure 1

6. $|-3.1| - |-4.2| + |2.3| =$

(A) 1.2

(B) 3.4

(C) 6.9

(D) 7.3

(E) 9.6

TURN TO THE NEXT PAGE.

LEVEL IC

KAPLAN PRACTICE TEST B

DO YOUR FIGURING HERE.

7. If $x \neq 0$, then $\dfrac{x+1}{6x} + \dfrac{x+1}{2x} =$

   (A) $\dfrac{2x+2}{3x}$

   (B) $\dfrac{2x+2}{4x}$

   (C) $\dfrac{2x+3}{6x}$

   (D) $\dfrac{2x+2}{8x}$

   (E) $\dfrac{x+2}{6x}$

8. If $8 = \dfrac{1}{\dfrac{1}{x^{-3}}}$, then $x =$

   (A) -2.00
   (B) -0.50
   (C) 0.50
   (D) 1.25
   (E) 2.00

9. Which of the following is the equation of a line that makes a 45-degree angle with the x-axis and has a y-intercept of 2 ?

   (A) $y = x + 2$
   (B) $y = x - 2$
   (C) $y = 45x + 2$
   (D) $y = 45x - 2$
   (E) $y = 2x + 45$

TURN TO THE NEXT PAGE.

KAPLAN PRACTICE TEST B

DO YOUR FIGURING HERE.

10. If x and y are both positive, x is even, and y is odd, which of the following must be odd?

   (A) xy

   (B) $x + 2y$

   (C) x^y

   (D) y^x

   (E) $\dfrac{x}{y}$

11. In Figure 2, if the length of AB is 2 more than three times the length of BC, and $AC = 14$, what is the length of BC ?

   (A) 2
   (B) 3
   (C) 4
   (D) 7
   (E) 12

Figure 2

12. Jean can paint a house in 10 hours and Dan can paint the same house in 12 hours. If Jean begins the job and does $\dfrac{1}{3}$ of it and then Dan takes over and finishes the job, what is the total time it takes them to paint the house?

   (A) 10 hours, 40 minutes
   (B) 11 hours
   (C) 11 hours, 20 minutes
   (D) 11 hours, 40 minutes
   (E) 12 hours

TURN TO THE NEXT PAGE.

LEVEL IC

KAPLAN PRACTICE TEST B

DO YOUR FIGURING HERE.

13. If $a = 4b$, $c = 8b^2$ and $b \neq 0$, then $\dfrac{c - a}{4b} =$

    (A) $-2b$

    (B) $1 - 2b$

    (C) $2b$

    (D) $2b - 1$

    (E) $2b + 3$

14. In Figure 3, if O is the center of the circle, and the ratio of x to y is 2 to 1, what is the ratio of a to b ?

    (A)   4:1

    (B)   3:1

    (C)   2:1

    (D)   1:2

    (E)   1:4

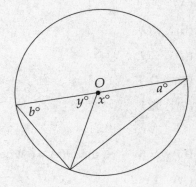

Figure 3

15. The cost of 2 sodas and 3 pretzels is $4.60. If the cost of 2 pretzels is $2.20, what is the cost of 2 sodas?

    (A) $0.60

    (B) $0.65

    (C) $1.20

    (D) $1.30

    (E) $2.40

TURN TO THE NEXT PAGE.

KAPLAN PRACTICE TEST B

DO YOUR FIGURING HERE.

16. If $\frac{2}{x}\left(x^2 + x\right) = \frac{1}{2}$, then $x + 1 =$

(A) $\frac{1}{4}$ (B) $\frac{1}{2}$ (C) 1 (D) $\frac{3}{2}$ (E) 2

17. In Figure 4, if 5 lines are drawn by connecting the labeled points with the origin, which of the lines would have the greatest slope?

(A) *AO*

(B) *BO*

(C) *CO*

(D) *DO*

(E) *EO*

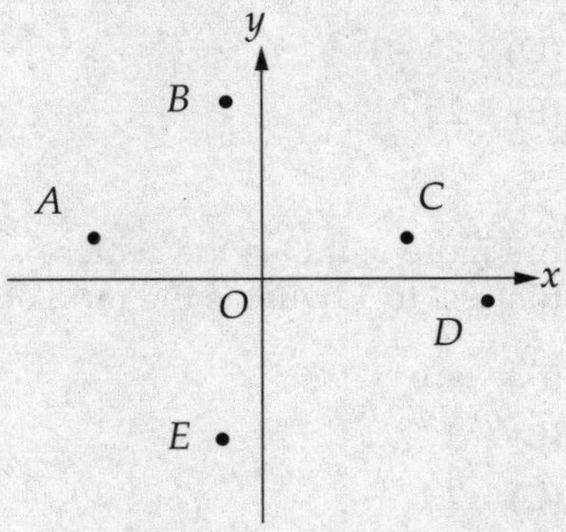

Figure 4

18. If $f(x) = 5x - x^2$, then $f(x) = 6$ when $x =$

(A) 1 only
(B) 2 only
(C) 1 or 5
(D) 2 or 3
(E) 2 or 6

19. If 25 percent of a certain number is 36, what would 40 percent of the same number be?

(A) 13.6
(B) 22.5
(C) 42.6
(D) 52.8
(E) 57.6

TURN TO THE NEXT PAGE.

DO YOUR FIGURING HERE.

20. In Figure 5, if the area of square *ABCD* is 5, what is the area of square *BEFD* ?

    (A)   7.07

    (B)   8.25

    (C)   10.00

    (D)   12.50

    (E)   25.00

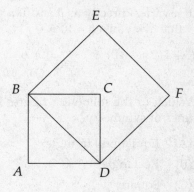

Figure 5

21. If $\sqrt{x} + 6 = x$, then $x =$

    (A)   4 only

    (B)   9 only

    (C)   4 or 9

    (D)   –4 or 9

    (E)   There is no solution.

22. A rubber ball is dropped from a height of 10 meters. If the ball always rebounds $\frac{4}{5}$ the distance it has fallen, how far, in meters, will the ball have traveled at the moment it hits the ground for the fourth time?

    (A)   4.10

    (B)   5.12

    (C)   29.52

    (D)   43.92

    (E)   49.04

TURN TO THE NEXT PAGE.

KAPLAN PRACTICE TEST B

DO YOUR FIGURING HERE.

23. If a varies directly as b and if $a = 4$ when $b = 5$, what is the value of a when $b = 10$?

(A) 4 (B) 5 (C) 6 (D) 8 (E) 9

24. Which of the following figures has the greatest number of lines of symmetry?

(A) Equilateral triangle
(B) Rectangle
(C) Square
(D) Rhombus
(E) Circle

25. What values for x would make $\frac{1}{\sqrt{x+1}}$ undefined?

(A) –1 only
(B) 1 only
(C) All real numbers greater than –1
(D) All real numbers less than –1
(E) All real numbers less than or equal to –1

TURN TO THE NEXT PAGE.

LEVEL IC

DO YOUR FIGURING HERE.

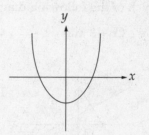

Figure 6

26. If the graph in Figure 6 represents $f(x)$, which of the following graphs would represent $|f(x)|$?

(A)

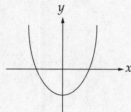

(B)

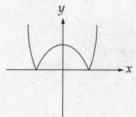

(C)

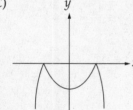

(D)

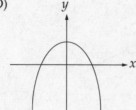

(E)

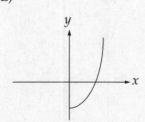

27. If $(x + y)^2 = (x - y)^2$, which of the following must be true?

   (A) $x = y$
   (B) $x = -y$
   (C) $x = 0$
   (D) $y = 0$
   (E) $x = 0$ or $y = 0$

TURN TO THE NEXT PAGE.

DO YOUR FIGURING HERE.

28. Which of the following diagrams represents the solution set

for $y \geq 2x + 3$ and $y \leq -\dfrac{1}{2}x + 1$?

(A)

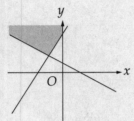

(B)

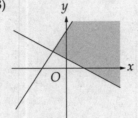

(C)

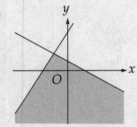

(D)

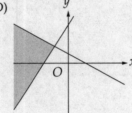

(E)

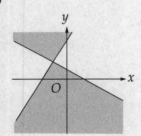

29. If $x \lozenge y = \sqrt{x^2 - y^2}$, then $10 \lozenge (-6) =$

(A) 2   (B) 8   (C) 16   (D) 32   (E) 64

TURN TO THE NEXT PAGE.

LEVEL IC

KAPLAN PRACTICE TEST B

30. In Figure 7, if *ABCDEF* is a regular hexagon, what is the length of *AE* if *AB* = 4 ?

    (A)   5.65
    (B)   5.73
    (C)   6.00
    (D)   6.93
    (E)   8.00

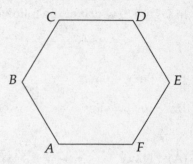

Figure 7

31. Which of the following is an equation of a line that will never intersect with the line that has the equation $y = 4x + 2$?

    (A) $y + 4x = 3$

    (B) $2y + 4x = 2$

    (C) $y - 4x = -3$

    (D) $y + 4x = -2$

    (E) $y + \dfrac{1}{4}x = 3$

32. If $i = \sqrt{-1}$, then $(3 + i)(3 - i) =$

    (A) 8   (B) 9   (C) 10   (D) $9 - i$ (E) $9 + i$

TURN TO THE NEXT PAGE.

259

LEVEL IC

KAPLAN PRACTICE TEST B

DO YOUR FIGURING HERE.

33. In Figure 8, which of the following must be true?

    I. $\sin x = \dfrac{3}{5}$

    II. $\tan y = \tan r$

    III. $\cos x = \sin s$

    (A) I only
    (B) III only
    (C) I and II only
    (D) I and III only
    (E) I, II, and III

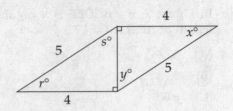

Figure 8

34. Figure 9 shows a semicircle that is the graph of the equation $y = \sqrt{6x - x^2}$. If the semicircle is rotated 360° about the x–axis, what is the volume of the sphere that is created?

    (A) 6π
    (B) 12π
    (C) 18π
    (D) 24π
    (E) 36π

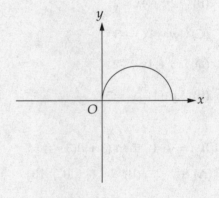

35. If $f(x) = 2x$ and $f(f(x)) = x + 1$, then $x =$

    (A) $\dfrac{1}{3}$ (B) 1   (C) 2   (D) 3   (E) 5

Figure 9

TURN TO THE NEXT PAGE.

LEVEL IC

KAPLAN PRACTICE TEST B

36. In Figure 10, if $\ell_1 \parallel \ell_2$, $x =$

    (A) 15
    (B) 25
    (C) 30
    (D) 35
    (E) 40

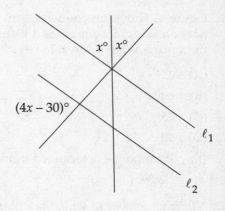

Figure 10

Note: Figure not drawn to scale.

37. Sally is interested in buying a car. If she has a choice of 3 colors (red, green, or blue), 2 body types (two or four doors) and 3 engine types (four, six, or eight cylinders), how many different models can she choose from?

    (A) 6   (B) 8   (C) 12   (D) 16   (E) 18

38. Which of the following is the solution set for $|2x| < 4$ and $|x - 2| < 2$?

    (A) $-2 < x < 0$
    (B) $-2 < x < 2$
    (C) $0 < x < 2$
    (D) $0 < x < 4$
    (E) $2 < x < 4$

39. If when $f(x)$ is divided by $3x + 1$, the quotient is $x^2 - x + 3$ and the remainder is 2, then $f(x) =$

    (A) $3x^3 - 2x^2 - 8x + 3$
    (B) $3x^3 - 2x^2 + 8x + 5$
    (C) $3x^3 + 4x^2 + 8x + 1$
    (D) $3x^3 - 4x^2 + 8x - 1$
    (E) $3x^3 - 4x^2 + 8x - 3$

TURN TO THE NEXT PAGE.

40. Figure 11 shows an equilateral triangle with two of its vertices on sides of a square and its third vertex on a vertex of the square. What is the value of $y - x$?

   (A)   45

   (B)   60

   (C)   75

   (D)   90

   (E)   It cannot be determined from the information given.

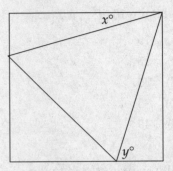

Figure 11

41. "If x is a member of set S, then x is <u>not</u> a member of set T" is logically equivalent to which of the following?

   I.   If x <u>is</u> a member of set T, then x is <u>not</u> a member of set S.

   II. If x is <u>not</u> a member of set T, then x <u>is</u> a member of set S.

   III. If x is <u>not</u> a member of set S, then x <u>is</u> a member of set T.

   (A)   I only

   (B)   II only

   (C)   III only

   (D)   I and II

   (E)   II and III

42. In Figure 12, if triangle ABC is an isosceles triangle of perimeter 20, what is the area of the circle with center O ?

   (A)   26.83

   (B)   33.65

   (C)   44.17

   (D)   57.33

   (E)   229.34

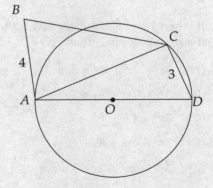

Figure 12

TURN TO THE NEXT PAGE.

LEVEL IC

DO YOUR FIGURING HERE.

43. If P is a point on the line $y = 2x$ in the first quadrant, and the distance from the origin to point P is 5, what are the coordinates of point P ?

    (A) (2.24, 4.47)

    (B) (3.00, 6.00)

    (C) (4.00, 8.00)

    (D) (4.47, 2.24)

    (E) (4.72, 2.36)

44. In Figure 13, circle O has diameter AB of length 8. If smaller circle P is tangent to diameter AB at point O and is also tangent to circle O, what is the area of the shaded region?

    (A)   3.14

    (B)   6.28

    (C)   9.42

    (D)   12.57

    (E)   25.13

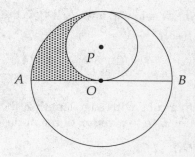

Figure 13

45. On a number line the coordinate of point A is 0, and the coordinate of point B is 6. If point P is located on the number line so that the distance from P to A is twice the distance from P to B, which of the following could be the coordinate of point P ?

    (A) 2 only

    (B) 4 only

    (C) 2 or 4

    (D) 4 or 12

    (E) 12 only

46. How many different-sized rectangular solids with a volume of 32 cubic units are there such that each dimension has an integer value?

    (A) 4   (B) 5   (C) 6   (D) 10   (E) 12

TURN TO THE NEXT PAGE.

DO YOUR FIGURING HERE.

47. If the roots of the equation $2x^2 + k = 8x$ are real, which of the following expresses all the possible values for k ?

(A) $k \leq 8$

(B) $k \geq 8$

(C) $k \geq 4$

(D) $k \leq -8$

(E) $k \geq -8$

48. If $f(x) = 1 - 4x$, and $f^{-1}(x)$ is the inverse of $f(x)$, then $f(-3)f^{-1}(-3) =$

(A) 1   (B) 3   (C) 4   (D) 10   (E) 13

49. In a cube with edge length 2, the distance from the center of one face to a vertex of the opposite face is

(A) 2.00

(B) 2.24

(C) 2.45

(D) 2.65

(E) 2.83

50. Set S is the set of all points (x, y) in the coordinate plane such that x and y are both integers with absolute value less than 4. If one of these points is chosen at random, what is the probability that this point will be 2 units or less from the origin?

(A) 0.189

(B) 0.227

(C) 0.265

(D) 0.314

(E) 0.356

STOP! END OF TEST. DO NOT TURN THE PAGE UNTIL YOU ARE READY TO CHECK YOUR ANSWERS.

Turn the page
for answers and explanations
to Level IC Test B.

Level IC Test B—Answer Key

1. D	11. B	21. B	31. C	41. A
2. A	12. C	22. E	32. C	42. D
3. B	13. D	23. D	33. D	43. A
4. E	14. D	24. E	34. E	44. B
5. C	15. D	25. E	35. A	45. D
6. A	16. A	26. B	36. D	46. B
7. A	17. E	27. E	37. E	47. A
8. C	18. D	28. D	38. C	48. E
9. A	19. E	29. B	39. B	49. C
10. D	20. C	30. D	40. B	50. C

1. (D)—For a to be as large as possible, make b as small as possible. If $b = 1$, then $a = 9$.

2. (A)—Isolate y:

$$2x + y = 2y - x$$
$$y - 2y = -x - 2x$$
$$-y = -3x$$
$$y = 3x$$

3. (B)—Set up a proportion:

$$\frac{240 \text{ miles}}{12 \text{ gallons}} = \frac{300 \text{ miles}}{x \text{ gallons}}$$
$$240x = 12 \times 300$$
$$x = 15$$

4. (E)—Solve for x:

$$2(1 + 2x) - 5(4 - 2x) = 14$$
$$2 + 4x - 20 + 10x = 14$$
$$14x - 18 = 14$$
$$14x = 32$$
$$x = \frac{32}{14} \approx 2.29$$

5. (C)—The angle marked 130° is an exterior angle and is equal to the sum of the remote interior angles marked $x°$ and $(x - 10)°$:

$$x + (x - 10) = 130$$
$$2x - 10 = 130$$
$$2x = 140$$
$$x = 70$$

6. (A)—Find the absolute values first, and then calculate:

$$|-3.1| - |-4.2| + |2.3| = 3.1 - 4.2 + 2.3$$
$$= -1.1 + 2.3$$
$$= 1.2$$

7. (A)—To add fractions, you need a common denominator. Here that would be $6x$. Multiply the top and bottom of the second fraction by 3 and you can proceed:

$$\frac{x+1}{6x} + \frac{x+1}{2x} = \frac{x+1}{6x} + \frac{3x+3}{6x}$$
$$= \frac{x+1+3x+3}{6x}$$
$$= \frac{4x+4}{6x}$$
$$= \frac{2x+2}{3x}$$

8. (C)—The right side of this equation is the reciprocal of the reciprocal of x^{-3}:

$$8 = \frac{1}{\frac{1}{x^{-3}}}$$

$$8 = x^{-3}$$

$$\frac{1}{x^3} = 8$$

$$x^3 = \frac{1}{8}$$

$$x = \frac{1}{2} = 0.50$$

9. (A)—To make a 45-degree angle with the x-axis, a line must have a slope of 1 or −1. Of the given choices, only (A) and (B) have such a slope, and of those only (A) has a y-intercept of 2.

10. (D)—(A) is always even because the product of an even and an odd is even. (B) is always even because the sum of an even and twice an odd is even. (C) is always even because no matter what positive integer exponent you raise an even number to, the result will be even. (D), on the other hand, is always odd because no matter what positive integer exponent you raise an odd number to, the result will be odd. (E) can never be odd because if an even divided by an odd gave you an integer at all, that integer would have to be even.

11. (B)—Mark up the figure:

You can see that:

$$(3x + 2) + x = 14$$

$$4x + 2 = 14$$

$$4x = 12$$

$$x = 3$$

12. (C)—If Jean can do the whole job in 10 hours, then it takes her $\frac{1}{3} \times 10 = 3\frac{1}{3}$ hours to do $\frac{1}{3}$ of the job. If Dan can do the whole job in 12 hours, then it takes him $\frac{2}{3} \times 12 = 8$ hours to do the other $\frac{2}{3}$ of the job. Thus, together, it takes them $3\frac{1}{3} \times 8 = 11\frac{1}{3}$ hours, or 11 hours and 20 minutes, to paint the house.

13. (D)—Plug $a = 4b$ and $c = 8b^2$ into the expression:

$$\frac{c-a}{4b} = \frac{8b^2 - 4b}{4b}$$

$$= \frac{4b(2b-1)}{4b}$$

$$= 2b - 1$$

14. (D)—Angles x and y are supplementary, so if $x + y = 180$ and $x = 2y$, then $x = 120$ and $y = 60$. Because all radii are equal, you can tell that the triangle with angles b and y is equilateral, so $b = 60$, and the triangle with angles a and x is isosceles with a vertex angle of 120°, so $a = 30$. The ratio of a to b, then, is 1:2.

15. (D)—If the cost of 2 pretzels is \$2.20, then each pretzel costs \$1.10, and 3 pretzels cost \$3.30. Two sodas, then, cost \$4.60 − \$3.30 = \$1.30.

16. (A)—Notice that if you distribute just the $\dfrac{1}{x}$, you'll get an equation in terms of $x + 1$:

$$\frac{2}{x}\left(x^2 + x\right) = \frac{1}{2}$$

$$2\left(\frac{x^2 + x}{x}\right) = \frac{1}{2}$$

$$2(x + 1) = \frac{1}{2}$$

$$x + 1 = \frac{1}{4}$$

17. (E)—The more "uphill" the line (as you move from left to right), the greater the slope. Lines *AO*, *BO*, and *DO* will all be "downhill" lines—that is, they all have negative slopes. Lines *CO* and *EO* will be uphill—they have positive slopes—but *EO* is visibly "steeper" than *CO*:

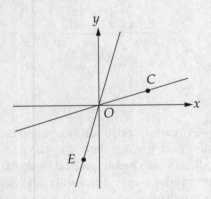

Line *EO* has the greatest slope.

18. (D)—If $f(x)$ equals both $5x - x^2$ and 6 for some particular value of x, then you can set $5x - x^2$ equal to 6 and solve for x:

$$5x - x^2 = 6$$

$$x^2 - 5x + 6 = 0$$

$$(x - 2)(x - 3) = 0$$

$$x = 2 \text{ or } 3$$

19. (E)—If 25% of a number is 36, then that number is 36 times 4, or 144. Forty percent of that is $(.40)(144) = 57.6$.

20. (C)—The area of the smaller square is 5, so each side is $\sqrt{5}$. Diagonal *BD* divides the square into 45–45–90 triangles, so the hypotenuse *BD* equals one of the legs times $\sqrt{2}$:

$$BD = \left(\sqrt{5}\right)\left(\sqrt{2}\right) = \sqrt{10}$$

Now that you know that one side of the big square is $\sqrt{10}$, you know that the area of the big square is $\left(\sqrt{10}\right)^2 = 10$.

21. (B)—When solving this equation algebraically, you will square both sides of the equation, which can result in a false solution:

$$\sqrt{x} + 6 = x$$
$$\sqrt{x} = x - 6$$
$$\left(\sqrt{x}\right)^2 = (x-6)^2$$
$$x = x^2 - 12x + 36$$
$$x^2 - 13x + 36 = 0$$
$$(x-4)(x-9) = 0$$
$$x = 4 \text{ or } 9$$

Try $x = 4$ and $x = 9$ in the original equation, and you'll find that $x = 4$ doesn't really work. The only solution is $x = 9$.

22. (E)—Sketch a diagram. First the ball drops 10 meters and rebounds $\frac{4}{5}$ of that, or 8 meters:

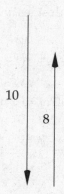

Then it drops 8 meters and rebounds $\frac{4}{5}$ of that, or 6.4 meters, and continuing this way you get:

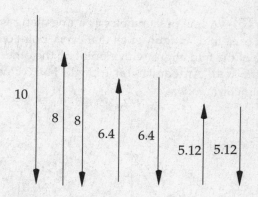

The sum of these 7 distances is 10 + 8 + 8 + 6.4 + 6.4 + 5.12 + 5.12 = 49.04.

23. (D)—If a varies directly as b, then a is equal to some constant times b:

$$a = kb$$
$$4 = k(5)$$
$$k = .8$$

Now plug in $b = 10$ and $k = .8$:

$$a = kb = (.8)(10) = 8$$

24. (E)—A line of symmetry is a line along which you can "fold" a figure such that every point on one side of the fold aligns with a point on the other side of the fold. An equilateral triangle has 3 lines of symmetry:

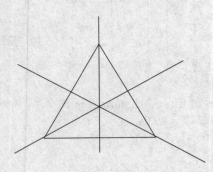

A nonsquare rectangle has 2, a nonsquare rhombus has 2, and a square has 4:

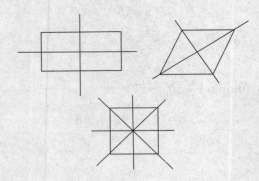

A circle, however, has infinitely many. Any fold through the center of the circle is a line of symmetry:

25. (E)—The expression will be undefined when the denominator is zero or when the expression under the radical is negative. The denominator is zero when:

$$\sqrt{x+1} = 0$$
$$x+1 = 0$$
$$x = -1$$

The expression under the radical is negative when:

$$x+1 < 0$$
$$x < -1$$

So the expression is undefined for all $x \leq -1$.

26. (B)—The graph of $|f(x)|$ is the same as the graph of $f(x)$ except that any part below the x-axis has to be flipped above the x-axis. That's what choice (B) shows.

27. (E)—Expand both sides and simplify:

$$(x+y)^2 = (x-y)^2$$
$$x^2 + 2xy + y^2 = x^2 - 2xy + y^2$$
$$2xy = -2xy$$
$$4xy = 0$$
$$xy = 0$$

Either $x = 0$ or $y = 0$.

28. (D)—The right sides of both inequalities are in $mx + b$ form, so it's easy to find the slopes and y-intercepts. In the first inequality the slope is 2 and the y-intercept is 3. In the second inequality the slope is $-\frac{1}{2}$—the negative reciprocal of the other slope, which makes the lines perpendicular—and the *y*-intercept is 1. All choices have the lines in the right places. The difference is in the shading. The first inequality has a "greater-than-or-equal-to" sign, so to satisfy that inequality a point must be on or above the line $y = 2x + 3$. The second inequality has a "less-than-or-equal-to" sign, so to satisfy that inequality a point must be on or below the line $y = -\dfrac{1}{2}x + 1$. The choice with the proper shading, then, is (D).

29. **(B)**—Plug $x = 10$ and $y = -6$ into the definition:

$$x \diamond y = \sqrt{x^2 - y^2}$$
$$10 \diamond (-6) = \sqrt{10^2 - (-6)^2} = \sqrt{100 - 36}$$
$$= \sqrt{64} = 8$$

30. **(D)**—Mark up the figure. Draw in not only the segment *AE* you're looking for, but also the perpendicular that makes the two right triangles as shown:

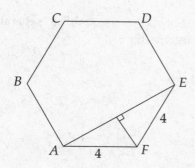

Each of the interior angles of a regular hexagon measures 120°, and these two right triangles split one of those 120° angles. That means the two right triangles are in fact 30-60-90s:

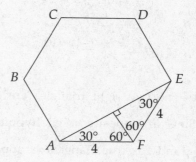

The hypotenuse of each 30–60–90 is 4, so the shared short leg is 2, and the long legs are each $2\sqrt{3}$. Segment *AE* is composed of two legs of $2\sqrt{3}$, so $AE = 4\sqrt{3} \approx 6.93$.

31. **(C)**—Never intersecting means parallel or, in other words, having the same slope. The slope of the equation in the question stem is 4. Put the answer choices into slope-intercept form until you find the one with a slope of 4:

(A) $y + 4x = 3$
           $y = -4x + 3$ Slope $= -4$

(B) $2y + 4x = 2$
           $2y = -4x + 2$
           $y = -2x + 1$ Slope $= -2$

(C) $y - 4x = -3$
           $y = 4x - 3$ Slope $= 4$

32. **(C)**—Use FOIL:

$$(3+i)(3-i) = 3 \cdot 3 + 3(-i) + i(3) - i^2$$
$$= 9 - 3i + 3i - i^2$$
$$= 9 - i^2$$
$$= 9 - (-1) = 10$$

33. **(D)**—The two right triangles are congruent 3-4-5s. Sine is opposite over hypotenuse, so $\sin x = \frac{3}{5}$, and I is true. Tangent is opposite over adjacent, so $\tan y = \frac{4}{3}$ and $\tan r = \frac{3}{4}$, and II is not true. Cosine is adjacent over hypotenuse and sine is opposite over hypotenuse, so $\cos x = \frac{4}{5}$ and $\sin s = \frac{4}{5}$, and III is true.

34. **(E)**—To find the volume of the sphere, you need the radius, which is the same as the radius of the semicircle. Find the x-intercepts—that is, the points at which $y = 0$:

$$\sqrt{6x - x^2} = 0$$
$$6x - x^2 = 0$$
$$x(6 - x) = 0$$
$$x = 0 \text{ or } 6$$

The distance between the x-intercept points is the diameter, so diameter = 6 and radius = 3. Now plug $r = 3$ into the sphere volume formula:

$$\text{Volume of sphere} = \frac{4}{3}\pi r^3 = \frac{4}{3}\pi(3)^3 = 36\pi$$

35. **(A)**—To perform the function on x means to double x. To perform the function twice would mean to double x and then to double the result—in effect turning x into $4x$. For some particular value of x, when you perform the function twice, you end up with $x + 1$. Therefore, for this value of x, $4x$ and $x + 1$ are the same:

$$4x = x + 1$$
$$3x = 1$$
$$x = \frac{1}{3}$$

36. **(D)**—Because ℓ_1 and ℓ_2 are parallel, you will need to look for transversals. Think of the two angles marked $x°$ as one angle measuring $2x°$. That angle and the angle marked $(4x - 30)°$ are exterior angles on the same side of the transversal, so they add up to $180°$:

$$2x + (4x - 30) = 180$$
$$6x - 30 = 180$$
$$6x = 210$$
$$x = 35$$

37. **(E)**—To get the total number of possibilities, multiply: 3 colors, 2 body types, and 3 engine types means $3 \times 2 \times 3 = 18$ different models.

38. **(C)**—Solve the inequalities separately, then find their intersection:

$$|2x| < 4$$
$$-4 < 2x < 4$$
$$-2 < x < 2$$

$$|x - 2| < 2$$
$$-2 < x < -2 < 2$$
$$0 < x < 4$$

The intersection of these ranges is $0 < x < 2$.

39. (B)—When the polynomial you're looking for is divided by $3x + 1$, the result is $x^2 - x + 3$, with a remainder of 2. To reconstruct the original polynomial, multiply $3x + 1$ by $x^2 - x + 3$ and add 2:

$$f(x) = (3x+1)(x^2 - x + 3) + 2$$
$$= 3x^3 - 3x^2 + 9x + x^2 - x + 3 + 2$$
$$= 3x^3 - 2x^2 + 8x + 5$$

40. (B)—Mark up the figure. Fill in the angles you know:

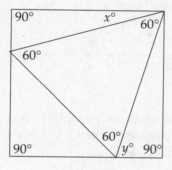

The right triangle at the top and the one on the right are congruent—they have the same hypotenuse and the same longer leg—so the two unknown angles in the upper right are equal:

$$2x + 60 = 90$$
$$2x = 30$$
$$x = 15$$

And now that you know the other two angles in that triangle with y, you can solve for y:

$$y + 15 + 90 = 180$$
$$y = 75$$

Therefore, $y - x = 75 - 15 = 60$.

41. (A)—Think of the original statement as, "If p, then q," in which p is "x is a member of set S," and q is "x is not a member of set T." "If p, then q" is equivalent to: "If not q then not p;" or: "If x is a member of set T, then x is not a member of set S." That's statement I. Statement II does not follow because it's: "If q, then p"; and statement III does not follow because it's "If not p, then not q."

42. (D)—The perimeter of triangle ABC is 20. One side is given as 4, so the other two sides add up to 16. The three sides cannot be 4, 4, and 12, because that violates the Triangle Inequality Theorem. The sides must be 4, 8, and 8:

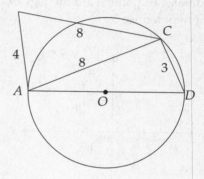

You know that triangle ACD is a right triangle because side AD is a diameter and C is a point on the circle. The legs of triangle ACD are 8 and 3, so the hypotenuse $AD = \sqrt{8^2 + 3^2} = \sqrt{73}$. The radius is half that, or $\dfrac{\sqrt{73}}{2}$. Plug $r = \dfrac{\sqrt{73}}{2}$ into the circle area formula:

$$\text{Area of Circle} = \pi \left(\frac{\sqrt{73}}{2} \right)^2 = \frac{73\pi}{4} \approx 57.33$$

43. (A)—Sketch a diagram:

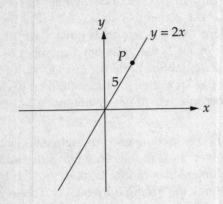

The coordinates of point P are the legs of the right triangle of hypotenuse 5:

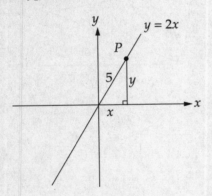

So $x^2 + y^2 = 5^2$. Furthermore, $y = 2x$, so:

$$x^2 + (2x)^2 = 25$$
$$5x^2 = 25$$
$$x^2 = 5$$
$$x = \sqrt{5} \approx 2.24$$
$$y = 2x = 2\sqrt{5} \approx 4.47$$

44. (B)—The radius of circle O is 4, so the area of circle O is $\pi(4)^2 = 16\pi$. The radius of circle P is 2, so the area of circle P is $\pi(2)^2 = 4\pi$. Think of the shaded region as the difference between a quarter of circle O and a half of circle P:

Shaded area $= \dfrac{1}{4}(16\pi) - \dfrac{1}{2}(4\pi)$
$= 4\pi - 2\pi$
$= 2\pi \approx 6.28$

45. (D)—Sketch a diagram:

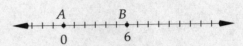

Now, where to put P so that it's twice as far from A as from B. One possibility is two-thirds of the way from A to B, that is, at 4:

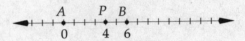

But that's not the only possibility. There's also the point that's the same distance to the right of B as B is from A, that is, at 12:

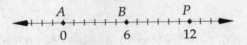

46. (B)—You're looking for sets of 3 positive integers with a product of 32. The easiest way to do that is just to list them systematically. If the smallest dimension is 1, then the product of the other two dimensions is 32, so there are these possibilities that include an edge of 1:

$$1 \times 1 \times 32$$
$$1 \times 2 \times 16$$
$$1 \times 4 \times 8$$

If the smallest dimension is 2, then the product of the other two dimensions is 16, so there are these additional possibilities:

$$2 \times 2 \times 8$$
$$2 \times 4 \times 4$$

You can't have a smallest dimension of 4, because then the product of the other two dimensions would have to be 8, and you can't make a product of 8 out of two integers greater than or equal to 4. So the five possibilities listed above are the only possibilities.

47. **(A)**—Reexpress the equation in standard form:

$$2x^2 + k = 8x$$
$$2x^2 - 8x + k = 0$$

Then use the quadratic formula:

$$x = \frac{-b \pm \sqrt{b^2 - 4ac}}{2a}$$
$$= \frac{8 \pm \sqrt{64 - 8k}}{4}$$

This expression will be real as long as what's under the radical is nonnegative:

$$64 - 8k \ge 0$$
$$-8k \ge -64$$
$$k \le 8$$

48. **(E)**—First find the inverse of $f(x)$:

$$f(x) = 1 - 4x$$
$$y = 1 - 4x$$
$$y + 4x = 1$$
$$4x = 1 - y$$
$$x = \frac{1 - y}{4}$$
$$f^{-1}(x) = \frac{1 - x}{4}$$

Now find $f(-3)$ and $f^{-1}(-3)$:

$$f(x) = 1 - 4x$$
$$f(-3) = 1 - 4(-3) = 1 - (-12) = 13$$

$$f^{-1}(x) = \frac{1 - x}{4}$$
$$f^{-1}(-3) = \frac{1 - (-3)}{4} = \frac{4}{4} = 1$$

And now, to get $f(-3)f^{-1}(-3)$, multiply:

$$f(-3)\,f^{-1}(-3) = 13(1) = 13$$

49. **(C)**—Sketch a diagram:

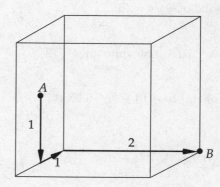

To get from A to B, you go down 1, back 1, and over 2. The straight-line distance, then, is:

$$d = \sqrt{1^2 + 1^2 + 2^2} = \sqrt{6} \approx 2.45$$

50. **(C)**—Probability is favorable outcomes over total outcomes. Here the total number of possible outcomes is the number of points with both coordinates of absolute integer value less than 4. That's all the points with x-coordinates of $-3, -2, -1, 0, 1, 2,$ or 3 and y-coordinates of $-3, -2, -1, 0, 1, 2,$ or 3. Seven possibilities for x and seven possibilities for y makes $7 \times 7 = 49$ possibilities for (x, y):

$$\text{Total outcomes} = 49$$

Next figure out how many of those points are 2 units or less from the origin. In other words, you're looking for points (x, y) such that $x^2 + y^2 \leq 4$. There are 9 points on the axes that work:

$$(0, -2), (0, -1), (0, 0), (0, 1), (0, 2), (-2, 0), (-1, 0),$$
$$(1, 0), (2, 0)$$

Additionally there are these four points that fit:

$$(-1, -1), (-1, 1), (1, -1), (1, 1)$$

That's a total of $9 + 4 = 13$:

$$\text{Favorable outcomes} = 13$$

So the probability is $13 \div 49 \approx 0.265$.

Step 1: Figure out your raw score. Refer to your answer sheet for the number right and the number wrong on the practice test you're scoring. (If you haven't checked your answers, do that now, using the answer key that follows the test.) You can use the chart below to figure out your raw score. Multiply the number wrong by .25 and subtract the result from the number right. Round the result to the nearest whole number. This is your raw score.

LEVEL IC TEST A

NUMBER RIGHT	NUMBER WRONG	RAW SCORE
☐	− (.25 × ☐)	= ☐ (ROUNDED)

LEVEL IC TEST B

NUMBER RIGHT	NUMBER WRONG	RAW SCORE
☐	− (.25 × ☐)	= ☐ (ROUNDED)

Step 2: Find your practice test score. Find your raw score in the left column of the table below. The score in the right column is your Level IC Practice Test score.

Find Your Practice Test Score

Raw	Scaled	Raw	Scaled	Raw	Scaled	Raw	Scaled	Raw	Scaled	Raw	Scaled
50	800	39	670	28	550	17	450	6	350	−5	260
49	790	38	660	27	540	16	450	5	350	−6	260
48	780	37	650	26	530	15	440	4	340	−7	250
47	770	37	640	25	520	14	430	3	330	−8	240
46	750	35	620	24	510	13	420	2	320	−9	230
45	740	34	610	23	500	12	400	1	310	−10	230
44	730	33	600	22	500	11	390	0	310	−11	220
43	720	32	590	21	490	10	390	−1	300	−12	210
42	700	31	580	20	480	9	380	−2	290		
41	690	30	570	19	470	8	370	−3	280		
40	680	29	560	18	460	7	360	−4	270		

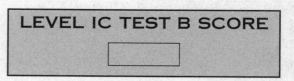

LEVEL IC TEST A SCORE

☐

LEVEL IC TEST B SCORE

☐

A note on your practice test scores: Don't take these scores too literally. Practice test conditions cannot precisely mirror real test conditions. Your actual SAT II: Mathematics Subject Test score will almost certainly vary from your practice test scores. Your scores on the practice tests will give you a rough idea of your range on the actual exam.

Level IIC
Practice Tests

This section contains two full-length Kaplan SAT II: Mathematics Level IIC Practice Tests. Answers and explanations appear after each test. Scoring information can be found in "Compute Your Level IIC Score" at the end of the section.

Kaplan's Level IIC Practice Test A

- The test that follows offers realistic practice for the SAT II: Mathematics Level IIC Test. To get the most out of it, you should take it under testlike conditions.

- Take the test in a quiet room with no distractions. Bring some No. 2 pencils and your calculator.

- Time yourself. You should spend no more than one hour on the 50 questions.

- Use the answer sheet to mark your answers.

- Answers and explanations follow the test.

- Scoring instructions are in "Compute your Level IIC Score" at the back of this section.

KAPLAN PRACTICE TEST A
ANSWER SHEET

1 Ⓐ Ⓑ Ⓒ Ⓓ Ⓔ	14 Ⓐ Ⓑ Ⓒ Ⓓ Ⓔ	27 Ⓐ Ⓑ Ⓒ Ⓓ Ⓔ	40 Ⓐ Ⓑ Ⓒ Ⓓ Ⓔ
2 Ⓐ Ⓑ Ⓒ Ⓓ Ⓔ	15 Ⓐ Ⓑ Ⓒ Ⓓ Ⓔ	28 Ⓐ Ⓑ Ⓒ Ⓓ Ⓔ	41 Ⓐ Ⓑ Ⓒ Ⓓ Ⓔ
3 Ⓐ Ⓑ Ⓒ Ⓓ Ⓔ	16 Ⓐ Ⓑ Ⓒ Ⓓ Ⓔ	29 Ⓐ Ⓑ Ⓒ Ⓓ Ⓔ	42 Ⓐ Ⓑ Ⓒ Ⓓ Ⓔ
4 Ⓐ Ⓑ Ⓒ Ⓓ Ⓔ	17 Ⓐ Ⓑ Ⓒ Ⓓ Ⓔ	30 Ⓐ Ⓑ Ⓒ Ⓓ Ⓔ	43 Ⓐ Ⓑ Ⓒ Ⓓ Ⓔ
5 Ⓐ Ⓑ Ⓒ Ⓓ Ⓔ	18 Ⓐ Ⓑ Ⓒ Ⓓ Ⓔ	31 Ⓐ Ⓑ Ⓒ Ⓓ Ⓔ	44 Ⓐ Ⓑ Ⓒ Ⓓ Ⓔ
6 Ⓐ Ⓑ Ⓒ Ⓓ Ⓔ	19 Ⓐ Ⓑ Ⓒ Ⓓ Ⓔ	32 Ⓐ Ⓑ Ⓒ Ⓓ Ⓔ	45 Ⓐ Ⓑ Ⓒ Ⓓ Ⓔ
7 Ⓐ Ⓑ Ⓒ Ⓓ Ⓔ	20 Ⓐ Ⓑ Ⓒ Ⓓ Ⓔ	33 Ⓐ Ⓑ Ⓒ Ⓓ Ⓔ	46 Ⓐ Ⓑ Ⓒ Ⓓ Ⓔ
8 Ⓐ Ⓑ Ⓒ Ⓓ Ⓔ	21 Ⓐ Ⓑ Ⓒ Ⓓ Ⓔ	34 Ⓐ Ⓑ Ⓒ Ⓓ Ⓔ	47 Ⓐ Ⓑ Ⓒ Ⓓ Ⓔ
9 Ⓐ Ⓑ Ⓒ Ⓓ Ⓔ	22 Ⓐ Ⓑ Ⓒ Ⓓ Ⓔ	35 Ⓐ Ⓑ Ⓒ Ⓓ Ⓔ	48 Ⓐ Ⓑ Ⓒ Ⓓ Ⓔ
10 Ⓐ Ⓑ Ⓒ Ⓓ Ⓔ	23 Ⓐ Ⓑ Ⓒ Ⓓ Ⓔ	36 Ⓐ Ⓑ Ⓒ Ⓓ Ⓔ	49 Ⓐ Ⓑ Ⓒ Ⓓ Ⓔ
11 Ⓐ Ⓑ Ⓒ Ⓓ Ⓔ	24 Ⓐ Ⓑ Ⓒ Ⓓ Ⓔ	37 Ⓐ Ⓑ Ⓒ Ⓓ Ⓔ	50 Ⓐ Ⓑ Ⓒ Ⓓ Ⓔ
12 Ⓐ Ⓑ Ⓒ Ⓓ Ⓔ	25 Ⓐ Ⓑ Ⓒ Ⓓ Ⓔ	38 Ⓐ Ⓑ Ⓒ Ⓓ Ⓔ	
13 Ⓐ Ⓑ Ⓒ Ⓓ Ⓔ	26 Ⓐ Ⓑ Ⓒ Ⓓ Ⓔ	39 Ⓐ Ⓑ Ⓒ Ⓓ Ⓔ	

right

wrong

Use the answer key following the test to count up the number of questions you got right and the number you got wrong. (Remember not to count omitted questions as wrong.) "Compute Your Level IIC Score" at the back of this section will show you how to find your score.

Remove this answer sheet and use it to complete Level IIC Test A.

LEVEL IIC

Practice Test A

50 Questions (1 hour)

Directions: For each question, choose the BEST answer from the choices given. If the precise answer is not among the choices, choose the one that best approximates the answer.

Notes:

(1) To answer some of these questions you will need a calculator. You must use at least a scientific calculator, but programmable and graphing calculators are also allowed.

(2) Figures in this test are drawn as accurately as possible UNLESS it is stated in a specific question that the figure is not drawn to scale. All figures are assumed to lie in a plane unless otherwise specified.

(3) The domain of any function f is assumed to be the set of all real numbers x for which $f(x)$ is a real number, unless otherwise indicated.

Reference Information: Use the following formulas as needed.

Right circular cone: If r = radius and h = height, then **Volume** $= \dfrac{1}{3}\pi r^2 h$; and if c = circumference of the base and ℓ = slant height, then **Lateral Area** $= \dfrac{1}{2}c\ell$.

Sphere: If r = radius, then **Volume** $= \dfrac{4}{3}\pi r^3$ and **Surface Area** $= 4\pi r^2$.

Pyramid: If B = area of the base and h = height, then **Volume** $= \dfrac{1}{3}Bh$.

DO YOUR FIGURING HERE.

1. If $x^3 = 7^5$, what is the value of x ?

   (A)   3.2
   (B)   11.6
   (C)   25.6
   (D)   243.0
   (E)   26,041.6

TURN TO THE NEXT PAGE.

LEVEL IIC

KAPLAN PRACTICE TEST A

DO YOUR FIGURING HERE.

2. If $a \, \Delta \, b \, \Delta \, c = \dfrac{ab}{c}$, which of the following equals 5 ?

   (A) $4 \, \Delta \, 3 \, \Delta \, 2$

   (B) $5 \, \Delta \, 2 \, \Delta \, 5$

   (C) $6 \, \Delta \, 4 \, \Delta \, 2$

   (D) $8 \, \Delta \, 4 \, \Delta \, 2$

   (E) $10 \, \Delta \, 2 \, \Delta \, 4$

3. If $f(x) = ex$ and $g(x) = \dfrac{x}{2}$, then $g(f(2)) =$

   (A) 2.7

   (B) 3.7

   (C) 4.2

   (D) 5.4

   (E) 6.1

4. If $\dfrac{x + 2y}{y} = 5$, what is the value of $\dfrac{y}{x}$?

   (A) –3   (B) $-\dfrac{1}{3}$ (C) $\dfrac{1}{3}$ (D) 3   (E) 4

5. In Figure 1, if $\cos \theta = 0.75$, $\tan \theta =$

   (A) 0.60

   (B) 0.67

   (C) 0.75

   (D) 0.88

   (E) 1.33

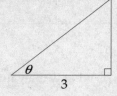

Figure 1

TURN TO THE NEXT PAGE.

KAPLAN PRACTICE TEST A

DO YOUR FIGURING HERE.

6. Which of the following is an equation of the line that has a y-intercept of 6 and an x-intercept of -2 ?

 (A) $3x - y = 6$

 (B) $3x - y = -6$

 (C) $3x + y = 6$

 (D) $6x + y = 3$

 (E) $6x - y = 3$

7. For all $y \neq 0$, $\dfrac{1}{y} + \dfrac{1}{2y} + \dfrac{1}{3y} =$

 (A) $\dfrac{1}{2y}$

 (B) $\dfrac{1}{6y}$

 (C) $\dfrac{5}{6y}$

 (D) $\dfrac{11}{6y}$

 (E) $\dfrac{1}{6y^3}$

8. In a class of 10 boys and 15 girls, the average score on a biology test is 90. If the average score for the girls is x, what is the average score for the boys in terms of x ?

 (A) $200 - \dfrac{2}{3}x$

 (B) $225 - \dfrac{3}{2}x$

 (C) $250 - 2x$

 (D) $250 - 3x$

 (E) $275 - 2x$

TURN TO THE NEXT PAGE.

DO YOUR FIGURING HERE.

9. Which of the following graphs is symmetric about the origin?

(A)

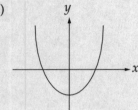

(B)

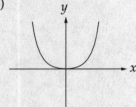

(C)

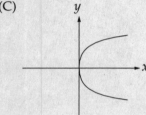

(D)

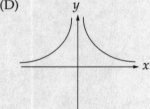

(E)

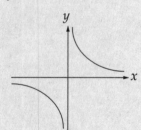

10. George is going on vacation and wishes to take along 2 books to read. If he has 5 different books to choose from, how many different combinations of 2 books can he bring?

(A) 2   (B) 5   (C) 10   (D) 15   (E) 20

TURN TO THE NEXT PAGE.

DO YOUR FIGURING HERE.

11. If $\sqrt{3-x} - x = 3$, $x =$

    (A) -1 or -6

    (B) 1 or -6

    (C) -1 only

    (D) -6 only

    (E)   There is no solution.

12. The linear equations $y = m_1x + 4$ and $y = m_2x + 3$ will intersect in the first or fourth quadrant if and only if

    (A) $m_1 = m_2$

    (B) $m_1 < m_2$

    (C) $m_1 > m_2$

    (D) $m_1 + m_2 = 0$

    (E) $m_1 \neq m_2$

13. If the probability that it will rain sometime on Monday is $\frac{1}{3}$ and the independent probability that it will rain sometime on Tuesday is $\frac{1}{2}$, what is the probability that it will rain on both days?

    (A) $\frac{1}{6}$ (B) $\frac{1}{5}$ (C) $\frac{1}{3}$ (D) $\frac{2}{5}$ (E) $\frac{5}{6}$

TURN TO THE NEXT PAGE.

LEVEL IIC

KAPLAN PRACTICE TEST A

DO YOUR FIGURING HERE.

14. If $\sin 2A = \dfrac{1}{2}$, then $\dfrac{1}{2 \sin A \cos A} =$

    (A) 1   (B) $\dfrac{3}{2}$ (C) 2   (D) 3   (E) 4

15. If $f(x) = x^2 - 4x + 1$, $f(x)$ crosses the x-axis at which of the following points?

    (A) $(-0.33, 0)$

    (B) $(0.27, 0)$

    (C) $(1.73, 0)$

    (D) $(3.27, 0)$

    (E) $(4.33, 0)$

16. If $-2 \le x \le 2$, the maximum value of $f(x) = 1 - x^2$ is

    (A) 2   (B) 1   (C) 0   (D) -1 (E) -2

17. For what values of x is $x^2 + 6 < 5x$?

    (A) $2 < x < 3$

    (B) $-2 < x < 3$

    (C) $-3 < x < -2$

    (D) $-2 < x$ or $x > 3$

    (E) $-3 < x$ or $x > 2$

18. Which of the following polynomials, when divided by $3x + 4$ equals $2x^2 + 5x - 3$ with remainder 3 ?

    (A) $6x^3 + 23x^2 - 11x - 12$

    (B) $6x^3 + 23x^2 - 11x - 9$

    (C) $6x^3 + 23x^2 - 11x - 15$

    (D) $6x^3 + 23x^2 + 11x - 12$

    (E) $6x^3 + 23x^2 + 11x - 9$

TURN TO THE NEXT PAGE.

LEVEL IIC

KAPLAN PRACTICE TEST A

DO YOUR FIGURING HERE.

19. Let $\lfloor x \rfloor$ be defined to be the "floor" of x, where $\lfloor x \rfloor$ is the greatest integer that is less than or equal to x, and let $\lceil x \rceil$ be the "ceiling" of x, where $\lceil x \rceil$ is the least integer that is greater than or equal to x. If $f(x) = \lceil x \rceil + \lfloor x \rfloor$ and x is not an integer, then $f(x)$ is also equal to

   (A) $2 \lceil x \rceil - 2$

   (B) $2 \lceil x \rceil$

   (C) $2 \lfloor x \rfloor$

   (D) $2 \lfloor x \rfloor + 1$

   (E) $2 \lfloor x \rfloor + 2$

20. If $\log_2 x + \log_2 x = 7$, then $x =$

   (A) 1.21

   (B) 1.40

   (C) 11.31

   (D) 18.52

   (E) 22.63

21. If $f(x) = \dfrac{\sqrt{x-1}}{x}$, what is the domain of $f(x)$?

   (A) All real numbers except for 0

   (B) All real numbers greater than or equal to 1

   (C) All real numbers less than or equal to 1

   (D) All real numbers greater than or equal to –1 but less than or equal to 1

   (E) All real number less than or equal to –1

TURN TO THE NEXT PAGE.

LEVEL IIC

KAPLAN PRACTICE TEST A

22. How many ways can 2 identical red chairs and 4 identical blue chairs be arranged in one row?

    (A) 6   (B) 15   (C) 21   (D) 24   (E) 30

23. If $a + b > 0$ and $c + d > 0$, which of the following must be true?

    (A) $a + b + c > 0$

    (B) $ac + bd > 0$

    (C) $a^2 + b^2 > 0$

    (D) $d(a + b) > 0$

    (E) $a + b > c + d$

24. If $x > 0$, $a = x \cos \theta$, and $b = x \sin \theta$, then $\sqrt{a^2 + b^2} =$

    (A) 1

    (B) x

    (C) $2x$

    (D) $x (\cos \theta + \sin \theta)$

    (E) $x \cos \theta \sin \theta$

25. If $0° < x < 90°$ and $5 \sin^2 x = 7 \sin x - 2$, what is the value of $\sin x$?

    (A) 1.00

    (B) 0.71

    (C) 0.40

    (D) 0.38

    (E) 0.35

TURN TO THE NEXT PAGE.

LEVEL IIC

KAPLAN PRACTICE TEST A

DO YOUR FIGURING HERE.

26. $3 - 2i$ and $3 + 2i$ are roots to which of the following quadratic equations?

    (A) $x^2 + 6x + 13 = 0$

    (B) $x^2 - 6x + 13 = 0$

    (C) $x^2 - 6x - 13 = 0$

    (D) $x^2 + 6x + 7 = 0$

    (E) $x^2 + 6x - 4 = 0$

27. In Figure 2, if point P is located on the unit circle, then $x + y =$

    (A) 0.37

    (B) 0.50

    (C) 0.78

    (D) 0.87

    (E) 1.37

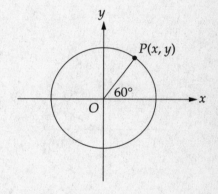

Figure 2

TURN TO THE NEXT PAGE.

DO YOUR FIGURING HERE.

28. If $0 \leq t \leq 1$, which of the following graphs is the graph of y versus x where x and y are related by the parametric equations $y = t^2$ and $x = \sqrt{t}$?

(A)
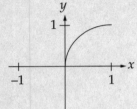

(B)
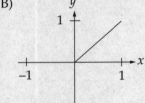

(C)
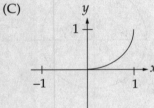

(D)
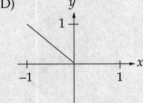

(E)
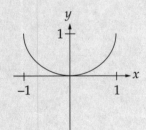

TURN TO THE NEXT PAGE.

DO YOUR FIGURING HERE.

29. In Figure 3, if isosceles right triangle ABC and square $ACDE$ share side AC, what is the degree measure of angle EBC ?

    (A) 27
    (B) 30
    (C) 60
    (D) 63
    (E) 75

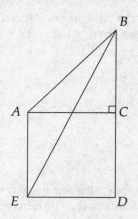

Figure 3

30. In Figure 4, which of the following denotes the correct vector arithmetic?

    (A) $\vec{x} + \vec{y} = \vec{z}$

    (B) $\vec{y} + \vec{z} = \vec{x}$

    (C) $\vec{x} + \vec{z} = \vec{y}$

    (D) $\vec{z} - \vec{x} = \vec{y}$

    (E) $\vec{z} - \vec{y} = \vec{x}$

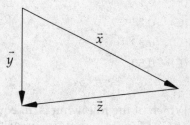

Figure 4

31. The horizontal distance, in feet, of a projectile that is fired with an initial velocity v, in feet per second, at an angle θ with the horizontal, is given by $H(v, \theta) = \dfrac{v^2 \sin(2\theta)}{32}$. If a football is kicked at an angle of 50 degrees with the horizontal and an initial velocity of 30 feet per second, what is the horizontal distance, in feet, from the point where the football is kicked to the point where the football first hits the ground?

    (A) 28   (B) 30   (C) 33   (D) 36   (E) 39

TURN TO THE NEXT PAGE.

32. If a right circular cone has a lateral surface area of 6π and a slant height 6, what is the radius of the base?

   (A) 0.50
   (B) 0.75
   (C) 1.00
   (D) 1.25
   (E) 1.50

33. If two fair dice are tossed, what is the probability that the two numbers that turn up are consecutive integers?

   (A) 0.14
   (B) 0.17
   (C) 0.28
   (D) 0.33
   (E) 0.50

34. Which of the following is an equation of the ellipse centered at $(-2, 3)$ with a minor axis of 4 parallel to the x-axis and a major axis of 6 parallel to the y-axis?

   (A) $\dfrac{(x-2)^2}{4} + \dfrac{(y-3)^2}{9} = 1$

   (B) $\dfrac{(x+2)^2}{4} + \dfrac{(y-3)^2}{9} = 1$

   (C) $\dfrac{(x-2)^2}{4} + \dfrac{(y+3)^2}{9} = 1$

   (D) $\dfrac{(x+2)^2}{4} + \dfrac{(y+3)^2}{9} = 1$

   (E) $\dfrac{(x-2)^2}{9} + \dfrac{(y+3)^2}{4} = 1$

TURN TO THE NEXT PAGE.

LEVEL IIC

KAPLAN PRACTICE TEST A

DO YOUR FIGURING HERE.

35. If $f(x) \geq 0$ and $g(x) \geq 0$ for all real x, which of the following statements must be true?

    I. $f(x) + g(x) \geq 0$

    II. $f(x) - g(x) \geq 0$

    III. $f(x)g(x) \geq 0$

    (A) I only

    (B) II only

    (C) III only

    (D) I and II

    (E) I and III

36. Where defined, $\dfrac{1 - \sin \theta}{1 - \csc \theta} =$

    (A) $\sin \theta$

    (B) $\csc \theta$

    (C) $-\sin \theta$

    (D) $-\csc \theta$

    (E) $-\cos \theta$

37. $\displaystyle\sum_{k=0}^{5} (-1)^k 2k =$

    (A) -10

    (B) -6

    (C) 0

    (D) 6

    (E) 10

TURN TO THE NEXT PAGE.

LEVEL IIC

KAPLAN PRACTICE TEST A

DO YOUR FIGURING HERE.

38. If $f(x) = x^3$, which of the following must be true?

(A) $f(-x) = f(x)$

(B) $f(-x) = -f(-x)$

(C) $f(-x) = -f(x)$

(D) $f(x) = -f(x)$

(E) $f(x) > f(-x)$

39. $\lim\limits_{x \to 1} \dfrac{x^2 - 6x + 5}{x^2 + 3x - 4} =$

(A) −1.25

(B) −0.80

(C) 0.80

(D) 1.25

(E) The limit does not exist.

40. If Figure 5 shows the graph of $f(x)$, what is the value of $f(f(3))$?

(A) −4

(B) −2

(C) 0

(D) 1

(E) 3

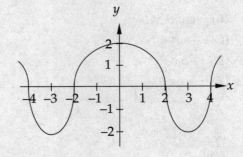

Figure 5

41. If all the terms of a geometric series are positive, the first term of the series is 2, and the third term is 8, how many digits are there in the 40th term?

(A) 10

(B) 11

(C) 12

(D) 13

(E) 14

TURN TO THE NEXT PAGE.

KAPLAN PRACTICE TEST A

DO YOUR FIGURING HERE.

42. In Figure 6, what is the degree measure, to the nearest integer, of angle *ABO* ?

(A) 50
(B) 48
(C) 45
(D) 43
(E) 40

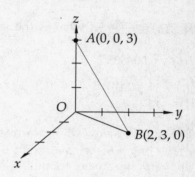

Figure 6

43. If $\log_2(x - 16) = \log_4(x - 4)$, which of the following could be the value of x ?

(A) 12   (B) 13   (C) 16   (D) 20   (E) 24

44. If a sphere of radius 3 is inscribed in a cube such that it is tangent to all six faces of the cube, the volume contained outside the sphere and inside the cube is

(A) 97   (B) 103   (C) 109   (D) 115   (E) 121

45. If $f(x) = \sin(\arctan x)$, $g(x) = \tan(\arcsin x)$, and $0 \leq x < \dfrac{\pi}{2}$, then $f\left(g\left(\dfrac{\pi}{10}\right)\right) =$

(A) 0.314
(B) 0.354
(C) 0.577
(D) 0.707
(E) 0.866

TURN TO THE NEXT PAGE.

DO YOUR FIGURING HERE.

46. If $f(x) = \dfrac{1}{(x+1)!}$, what is the smallest integer x such that

    $f(x) < 0.000005$?

    (A) 7   (B) 8   (C) 9   (D) 10   (E) 11

47. In Figure 7, point O has coordinates $(0, 0)$, point P lies on the graph of $y = 6 - x^2$, and point B has coordinates $(2\sqrt{3}, 0)$. If $OP = BP$, the area of triangle OPB is

    (A)   1.7
    (B)   3.0
    (C)   3.5
    (D)   4.7
    (E)   5.2

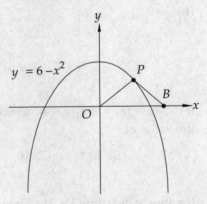

Figure 7

48. If $\cos 2x = \sin x$, and x is in radians, which of the following is a possible value of x ?

    (A)   0.39
    (B)   0.52
    (C)   1.05
    (D)   1.60
    (E)   2.09

TURN TO THE NEXT PAGE.

LEVEL IIC

KAPLAN PRACTICE TEST A

DO YOUR FIGURING HERE.

49. In Figure 8, if a wooden right circular cylinder with radius 2 meters and height 6 meters has a cylindrical hole of diameter 2 meters drilled through the center as shown, what is the entire surface area (including the top and bottom faces), in square meters, of the resulting figure?

(A) 38π

(B) 40π

(C) 42π

(D) 44π

(E) 46π

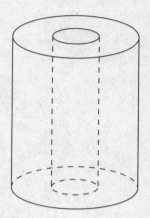

Figure 8

50. What is the greatest possible number of points of intersection between a parabola and a circle?

(A) 2   (B) 3   (C) 4   (D) 6   (E) 8

STOP! END OF TEST. DO NOT TURN THE PAGE UNTIL YOU ARE READY TO CHECK YOUR ANSWERS.

Level IIC Test A—Answer Key

1. C	11. C	21. B	31. A	41. D
2. E	12. B	22. B	32. C	42. E
3. B	13. A	23. C	33. C	43. D
4. C	14. C	24. B	34. B	44. B
5. D	15. B	25. C	35. E	45. A
6. B	16. B	26. B	36. C	46. B
7. D	17. A	27. E	37. B	47. E
8. B	18. E	28. C	38. C	48. B
9. E	19. D	29. A	39. B	49. C
10. C	20. C	30. C	40. C	50. C

1. **(C)**—Use your calculator. First find that $7^5 = 16,807$. Then find that the cube root of 16,807 is about 25.6.

2. **(E)**—Go down the answer choices and try multiplying the first two numbers and dividing by the third until you find the choice that yields 5. The answer is (E) because $\frac{10 \times 2}{4} = 5$.

3. **(B)**—Perform the inside function first:

$$f(x) = e^x$$
$$f(2) = e^2 \approx 7.389$$

Then perform the outside function on the result:

$$g(x) = \frac{x}{2}$$
$$g(7.389) = \frac{7.389}{2} \approx 3.7$$

4. **(C)**—Manipulate the equation to get $\frac{y}{x}$ on one side:

$$\frac{x+2y}{y} = 5$$
$$x + 2y = 5y$$
$$x = 3y$$
$$1 = \frac{3y}{x}$$
$$\frac{y}{x} = \frac{1}{3}$$

5. **(D)**—First use your calculator to solve for θ:

$$\cos\theta = .75$$
$$\theta = \arccos(.75) \approx 41.4°$$

Then find the tangent of the result:

$$\tan(41.4°) \approx 0.88$$

6. **(B)**—A y-intercept of 6 means that one point on the line is $(0, 6)$. Plug $x = 0$ and $y = 6$ into the answer choices and you'll find that only (B) and (C) work. Now test those two choices with the x-intercept $(-2, 0)$. Of (B) and (C), only (B) works this time.

7. **(D)**—To add fractions you need a common denominator. Here the LCD is $6y$:

$$\frac{1}{y}+\frac{1}{2y}+\frac{1}{3y}=\frac{6}{6y}+\frac{3}{6y}+\frac{2}{6y}$$
$$=\frac{11}{6y}$$

8. **(B)**—Let y represent the average score for the boys. Fifteen girls average x, so their 15 scores add up to $15x$. Ten boys average y, so their 10 scores add up to $10y$. The 25 students average 90, so their scores add up to $25 \times 90 = 2{,}250$. Now you can set up an equation and solve for y. The girls' total and the boys' total add up to the grand total of 2,250:

$$15x + 10y = 2{,}250$$
$$10y = 2{,}250 - 15x$$
$$y = \frac{2{,}250 - 15x}{10} = 225 - \frac{3}{2}x$$

9. **(E)**—To be symmetric about the origin means that, for any point A on the graph, there is another point B on the graph such that the origin is the midpoint of AB. So you should be able to start at any point on the graph, draw a straight line segment to the origin, continue straight the same distance beyond the origin, and you should be at another point on the graph. Thus, for example, you can see that (A) does not work:

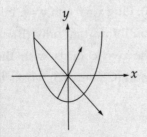

But (E) does work:

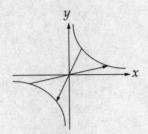

10. **(C)**—George has 5 choices for the first book, and then 4 choices for the second. That's 20 permutations for taking 2 books out of 5. But this is a combinations questions. Order doesn't matter. Whether he reads A or B first, it's the same combination of 2 books, so you have to divide the 20 permutations by 2. If you like formulas, here's how to do this one:

$$_nC_r = \frac{n!}{r!(n-r)!}$$
$$_5C_2 = \frac{5!}{2!3!} = 10$$

11. **(C)**—One of the steps in solving this equation is to square both sides. That step can result in "extraneous" solutions. Look:

$$\sqrt{3-x} - x = 3$$
$$\sqrt{3-x} = 3 + x$$
$$\left(\sqrt{3-x}\right)^2 = (3+x)^2$$
$$3 - x = 9 + 6x + x^2$$
$$x^2 + 7x + 6 = 0$$
$$(x+1)(x+6) = 0$$
$$x = -1 \text{ or } -6$$

But if you plug those solutions back into the original equation, you'll find that $x = -6$ doesn't work. The only solution is $x = -1$.

12. **(B)**—You can think this one through conceptually: The first equation intercepts the y-axis at $(0, 4)$, and the second equation intercepts the y-axis at $(0, 3)$. That they intersect in the first or fourth quadrant means that they meet up somewhere to the right of the y-axis:

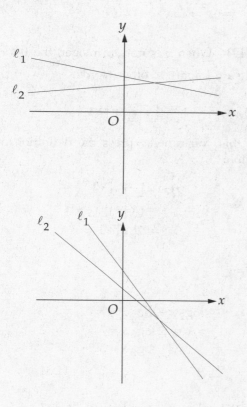

In both of the cases shown, ℓ_2 has the greater slope, so $m_2 > m_1$.

Alternatively, you could do this one algebraically. The point of intersection is the point at which $m_1x + 4 = m_2x + 3$:

$$m_1x + 4 = m_2x + 3$$
$$m_2x - m_1x = 4 - 3$$
$$x(m_2 - m_1) = 1$$
$$m_2 - m_1 = \frac{1}{x}$$

In the first and fourth quadrants, x is positive, and so therefore is $\frac{1}{x}$:

$$m_2 - m_1 = \frac{1}{x}$$
$$m_2 - m_1 > 0$$
$$m_2 > m_1$$

13. **(A)**—The combined probability of independent events is the product of the separate probabilities: $\frac{1}{3} \times \frac{1}{2} = \frac{1}{6}$.

14. **(C)**—You can use your calculator. Arcsin $\frac{1}{2}$ = $30°$, so $2A = 30°$ and $A = 15°$. Now use your calculator to find that $\frac{1}{2\sin 15° \cos 15°} = 2$.

But the solution's even quicker if you remember the double-angle sine formula:

$$\sin 2A = 2\sin A \cos A$$

Therefore:

$$\frac{1}{2\sin A \cos A} = \frac{1}{\sin 2A} = \frac{1}{0.5} = 2$$

15. (B)—The graph crosses the x-axis at the point where $f(x) = 0$. So:

$$x^2 - 4x + 1 = 0$$
$$x = \frac{4 \pm \sqrt{16 - 4}}{2}$$
$$= \frac{4 \pm \sqrt{12}}{2}$$
$$= 2 \pm \sqrt{3} \approx 3.73 \text{ or } 0.27$$

Choice (B) matches one of these.

16. (B)—You can think this one through conceptually. The expression $1 - x^2$ will be at its maximum when the x^2 that's subtracted from the 1 is as small as it can be. The square of a real number can't be any smaller than 0—and $x = 0$ is within the specified domain—so:

$$\text{maximum} = 1 - 0^2 = 1$$

17. (A)—First put the inequality into standard form:

$$x^2 + 6 < 5x$$
$$x^2 - 5x + 6 < 0$$

Then factor:

$$(x - 2)(x - 3) < 0$$

This product will be negative only when the smaller factor $x - 3$ is negative and the larger factor $x - 2$ is positive. That's when $2 < x < 3$.

18. (E)—To find the original polynomial, multiply $3x + 4$ by $2x^2 + 5x - 3$, and then add the remainder 3 to the result:

$$(3x + 4)(2x^2 + 5x - 3) + 3$$
$$= 6x^3 + 15x^2 - 9x + 8x^2 + 20x - 12 + 3$$
$$= 6x^3 + 23x^2 + 11x - 9$$

19. (D)—When x is not an integer, the floor and ceiling are 1 apart. In other words:

$$\lceil x \rceil = \lfloor x \rfloor + 1$$

With this you can reexpress the definition of the function:

$$f(x) = \lfloor x \rfloor + \lceil x \rceil$$
$$= \lfloor x \rfloor + (\lfloor x \rfloor + 1)$$
$$= 2\lfloor x \rfloor + 1$$

20. (C)—

$$\log_2 x + \log_2 x = 7$$
$$2\log_2 x = 7$$
$$\log_2 x = 3.5$$
$$x = 2^{3.5} \approx 11.31$$

21. (B)—To be in the domain of this function, x must not be anything that makes the expression under the radical negative or that makes the denominator zero. The expression under the radical is $x - 1$, and it must be nonnegative:

$$x - 1 \geq 0$$
$$x \geq 1$$

The denominator's simply x, which then cannot be zero. It's already been established, however, that x must be greater than or equal to 1, so that's the domain.

22. (B)—There is a formula that applies to this situation. The number of distinct permutations of n things, a of which are indistinguishable, b of which are indistinguishable, etcetera, is:

$$\frac{n!}{a!\,b!\cdots}$$

Here there are 6 chairs, 2 of which are indistinguishable and 4 of which are indistinguishable, so the number of permutations is:

$$\frac{6!}{2!\,4!} = \frac{6\cdot5\cdot4\cdot3\cdot2\cdot1}{2\cdot1\cdot4\cdot3\cdot2\cdot1} = \frac{6\cdot5}{2} = 15$$

23. (C)—The best way to go about this one is to check out each answer choice, trying to think of a case where that choice is not true. The correct answer is the one that has no counterexample. That $a + b > 0$ and $c + d > 0$ would imply, for example, that the total sum $a + b + c + d$ would also be positive, but that's not the same as saying (A), $a + b + c > 0$. If $a = 3$, $b = -2$, $c = -4$, and $d = 5$, then (A) is not true. Nor is (B). (C) is true for this set of numbers and for any possible set of numbers because $a^2 + b^2$ will be greater than zero as long as a and b are not both zero.

24. (B)—

$$\sqrt{a^2 + b^2} = \sqrt{(x\cos\theta)^2 + (x\sin\theta)^2}$$
$$= \sqrt{x^2(\cos^2\theta + \sin^2\theta)}$$
$$= \sqrt{x^2}$$
$$= |x|$$

It's given that $x > 0$, so $|x| = x$.

25. (C)—What you have here is a quadratic equation in which the unknown is $\sin x$. To make things simpler, replace $\sin x$ with y and solve for y:

$$5\sin^2 x = 7\sin x - 2$$
$$5y^2 = 7y - 2$$
$$5y^2 - 7y + 2 = 0$$
$$y = \frac{7 \pm \sqrt{49 - 40}}{10}$$
$$= \frac{7 \pm 3}{10}$$
$$= \frac{4}{10} \text{ or } \frac{10}{10}$$
$$= 0.40 \text{ or } 1$$

It's given that x is a positive acute angle, so $0 < \sin x < 1$, and only 0.40 fits.

26. (B)—If the solutions to the equation $ax^2 + bx + c = 0$ are $3 \pm 2i$, then:

$$\frac{-b}{2a} = 3 \quad \text{and} \quad \frac{\sqrt{b^2 - 4ac}}{2a} = 2i$$

In all the answer choices, $a = 1$, so you can say more simply:

$$\frac{-b}{2} = 3 \qquad \frac{\sqrt{b^2 - 4c}}{2} = 2i$$
$$b = -6 \qquad \sqrt{b^2 - 4c} = 4i$$
$$b^2 - 4c = -16$$
$$36 - 4c = -16$$
$$-4c = -52$$
$$c = 13$$

So the answer is the equation with $a = 1$, $b = -6$, and $c = 13$: $x^2 - 6x + 13 = 0$.

27. **(E)**—Make a right triangle:

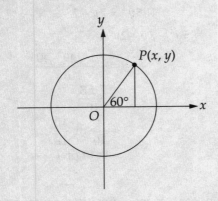

That's a 30-60-90 triangle. The hypotenuse is the radius, so it's 1. That means that the short leg is $\frac{1}{2}$ and the long leg is $\frac{\sqrt{3}}{2}$:

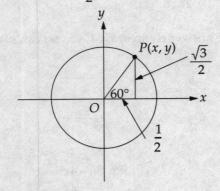

So, $x + y = \frac{1}{2} + \frac{\sqrt{3}}{2} \approx 1.37$.

28. **(C)**—Combine the equations so as to lose t and get y in terms of x:

$$x = \sqrt{t}$$
$$x^2 = t$$

$$y = t^2$$
$$= \left(x^2\right)^2 = x^4$$

So you might be tempted by (E), which looks like the graph of $y = x^4$. But the stem says that $0 \le t \le 1$, so the only possible values of $x = \sqrt{t}$ are $0 \le x \le 1$, so the correct graph is (C).

29. **(A)**—Mark up the figure. Call the sides of the square and the legs of the triangle each 1. What you're looking for is the measure of the angle marked $x°$ in right triangle BDE:

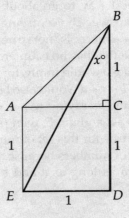

The leg opposite x is 1, and the leg adjacent to x is 2, so:

$$\tan x = \frac{\text{opposite}}{\text{adjacent}} = \frac{1}{2}$$

$$x = \arctan \frac{1}{2} \approx 27$$

30. **(C)**—The figure shows the head of $\vec{x}$ touching the tail of $\vec{z}$, so those are the two being added. The result is $\vec{y}$ because it then connects the tail of $\vec{x}$ to the head of $\vec{z}$. So:

$$\vec{x} + \vec{z} = \vec{y}$$

31. **(A)**—Just plug $\theta = 50$ and $v = 30$ into the formula and crank out the answer:

$$\begin{aligned} H &= \frac{v^2 \sin(2\theta)}{32} \\ &= \frac{\left(30^2\right)\sin(2 \times 50°)}{32} \\ &= \frac{900 \sin 100°}{32} \\ &\approx 28 \end{aligned}$$

32. **(C)**—A formula for the lateral area of a cone is given in the directions:

$$\text{Lateral Area} = \frac{1}{2}c\ell$$

Here the lateral area is 6π and $\ell = 6$, so you can solve for c:

$$\begin{aligned} 6\pi &= \frac{1}{2}c(6) \\ \pi &= \frac{1}{2}c \\ c &= 2\pi \end{aligned}$$

Now you can use the base circumference $c = 2\pi$ to find the base radius:

$$\begin{aligned} \text{Circumference} &= 2\pi r \\ 2\pi &= 2\pi r \\ r &= 1 \end{aligned}$$

33. **(C)**—The total number of possible outcomes is $6 \times 6 = 36$. Of those outcomes, the following are consecutive integers:

1 and 2
2 and 1
2 and 3
3 and 2
3 and 4
4 and 3
4 and 5
5 and 4
5 and 6
6 and 5

That's 10 favorable outcomes:

$$\begin{aligned} \text{Probability} &= \frac{\text{Favorable outcomes}}{\text{Total possible outcomes}} \\ &= \frac{10}{36} \approx 0.28 \end{aligned}$$

34. **(B)**—The equation of an ellipse centered at the point (p, q) and with axes $2a$ and $2b$ is:

$$\frac{\left(x-p\right)^2}{a^2} + \frac{\left(y-q\right)^2}{b^2} = 1$$

Here $p = -2$, $q = 3$, $a = 2$, and $b = 3$, so the equation is:

$$\frac{\left(x+2\right)^2}{2^2} + \frac{\left(y-3\right)^2}{3^2} = 1$$

which is the same as choice (B).

35. (E)—Don't let the functions symbolism confuse you. Just think of it as two quantities—$f(x)$ and $g(x)$—that are both nonnegative. Statement I says their sum is nonnegative—that's true. Add any two nonnegatives and you'll get a nonnegative sum. Statement II says the difference $f(x) - g(x)$ is nonnegative. Well, that's true only if $f(x) \geq g(x)$. But there's no reason that $g(x)$ couldn't be greater than $f(x)$. Statement III says their product is nonnegative—that's true. Multiply any two nonnegatives and you'll get a nonnegative product. Statements I and III are true.

36. (C)—Reexpress cosecant as 1 over sine:

$$\frac{1 - \sin\theta}{1 - \csc\theta} = \frac{1 - \sin\theta}{1 - \frac{1}{\sin\theta}}$$

$$= \frac{1 - \sin\theta}{\frac{\sin\theta}{\sin\theta} - \frac{1}{\sin\theta}}$$

$$= \frac{1 - \sin\theta}{\frac{\sin\theta - 1}{\sin\theta}}$$

$$= \frac{(1 - \sin\theta)(\sin\theta)}{\sin\theta - 1}$$

$$= \frac{-(\sin\theta - 1)(\sin\theta)}{\sin\theta - 1}$$

$$= -\sin\theta$$

37. (B)—Just plug in the six possible values for k, compute the results, and add them up:

$$k = 0 \Rightarrow (-1)^0 2(0) = 0$$
$$k = 1 \Rightarrow (-1)^1 2(1) = -2$$
$$k = 2 \Rightarrow (-1)^2 2(2) = 4$$
$$k = 3 \Rightarrow (-1)^3 2(3) = -6$$
$$k = 4 \Rightarrow (-1)^4 2(4) = 8$$
$$k = 5 \Rightarrow (-1)^5 2(5) = \underline{-10}$$
$$12 - 18 = -6$$

38. (C)—The function is cubing. Think about each answer choice. (A): The cube of minus x equals the cube of x? No. (B): The cube of minus x equals the opposite of the cube of minus x? No. (C): The cube of minus x equals the opposite of the cube of x? That sounds plausible. In fact, (C) is true. It doesn't matter whether you take the opposite of a number first and then cube it, or cube the number first and then take its opposite—you'll get the same result both ways.

39. (B)—The first step in finding a limit is generally to factor:

$$\frac{x^2 - 6x + 5}{x^2 + 3x - 4} = \frac{(x - 5)(x - 1)}{(x + 4)(x - 1)}$$

If x ever actually gets to 1, the expression becomes undefined—zero over zero. But if you cancel the $(x - 1)$ from the top and bottom:

$$\frac{(x - 5)(x - 1)}{(x + 4)(x - 1)} = \frac{x - 5}{x + 4}$$

you can plug in $x = 1$ and find the limit:

$$\lim_{x \to 1} \frac{x - 5}{x + 4} = \frac{1 - 5}{1 + 4}$$

$$= \frac{-4}{5} = -0.80$$

40. (C)—Don't try to figure out an equation to fit this weird graph. Just read the values right off the graph. First, find $f(3)$. Go to +3 on the x-axis and see what y is there. It's –2. Now find $f(-2)$. Go to –2 on the x-axis and see what y is there. It's 0:

$$f(3) = -2$$
$$f(f(3)) = f(-2) = 0$$

41. (D)—The first term is 2^1 and the third term is 2^3, so the 40th term is 2^{40}. Use your calculator and you'll get an answer in scientific notation something like:

$$1.0995 \text{ E12}$$

That is:

$$1.0995 \times 10^{12}$$

That's 1 followed by 12 digits, for a total of 13 digits.

42. (E)—The length of OA is 3, and the length of OB is $\sqrt{2^2 + 3^2} = \sqrt{13}$. OA over OB is the tangent of the angle you're looking for:

$$\tan x = \frac{OA}{OB} = \frac{3}{\sqrt{13}}$$
$$x = \arctan\left(\frac{3}{\sqrt{13}}\right) \approx 40°$$

43. (D)—Put everything in terms of $\log_2$:

$$\log_2(x-16) = \log_4(x-4)$$
$$\log_2(x-16) = \frac{\log_2(x-4)}{\log_2 4}$$
$$\log_2(x-16) = \frac{\log_2(x-4)}{2}$$
$$2\log_2(x-16) = \log_2(x-4)$$
$$\log_2(x-16)^2 = \log_2(x-4)$$
$$(x-16)^2 = x-4$$
$$x^2 - 32x + 256 = x - 4$$
$$x^2 - 33x + 260 = 0$$
$$(x-13)(x-20) = 0$$
$$x = 13 \text{ or } 20$$

Of those two apparent solutions, one is impossible. The log of a negative number is undefined, so $x = 13$ is an extraneous solution: you can't take the log of $(13 - 16)$. The only solution is 20.

44. (B)—The cube is $6 \times 6 \times 6$, so its volume is 216. The sphere has radius 3, so:

$$\text{Volume of sphere} = \frac{4}{3}\pi r^3$$
$$= \frac{4}{3}\pi(3^3)$$
$$= 36\pi$$

The difference is $216 - 36\pi \approx 103$.

45. (A)—You could use your calculator and do this one step by step. Set your calculator to radian mode. First perform the inside function:

$$g\left(\frac{\pi}{10}\right) = \tan\left(\arcsin\left(\frac{\pi}{10}\right)\right)$$
$$\approx \tan(0.3196)$$
$$\approx 0.3309$$

Then perform the outside function on the result:

$$f(0.3309) = \sin\left(\arctan(0.3309)\right)$$
$$\approx \sin(0.3196)$$
$$\approx 0.314$$

Far quicker and simpler would be to realize that if you take the sin of the arctan of the tan of the arcsin, you'll end up back where you started. The answer to this question is just the decimal approximation of the fraction $\frac{\pi}{10}$.

46. (B)—First convert 0.000005 into a fraction:

$$0.000005 = \frac{5}{1,000,000} = \frac{1}{200,000}$$

You're looking for the smallest integer value of x that will make $\frac{1}{(x+1)!}$ less than $\frac{1}{200,000}$. In other words, you're looking for the smallest integer x that will make $(x + 1)!$ greater than 200,000. Use your calculator and try a few possibilities. $1 \times 2 \times 3 \times 4 \times 5 \times 6 \times 7 \times 8 = 40,320$. Not big enough. But multiply that by 9 and you're up to 362,880. So $9! > 200,000$, $x + 1 = 9$, and therefore $x = 8$.

47. (E)—To find the area of the triangle, you want the base and the height. The base is the length OB, which is simply the x-coordinate of point B: $2\sqrt{3}$. The height is the y-coordinate of point P, which is equal to $6 - x^2$. The triangle is isosceles, so the altitude from P to base OB divides the base in half and the x-coordinate for point P is $\sqrt{3}$. Plug $x = \sqrt{3}$ into the equation $y = 6 - x^2$ to find the height:

$$\text{height} = 6 - \left(\sqrt{3}\right)^2 = 6 - 3 = 3$$

So if the base is $2\sqrt{3}$ and the height is 3:

$$\text{Area} = \frac{1}{2}(\text{base})(\text{height})$$
$$= \frac{1}{2}\left(2\sqrt{3}\right)(3) = 3\sqrt{3} \approx 5.2$$

48. (B)—Use the relationship $\cos 2x = 1 - 2\sin^2 x$ to get everything in terms of sine. And be sure your calculator is in radian mode.

$$\cos 2x = \sin x$$
$$1 - 2\sin^2 x = \sin x$$
$$2\sin^2 x + \sin x - 1 = 0$$
$$(2\sin x - 1)(\sin x + 1) = 0$$
$$2\sin x - 1 = 0 \text{ or } \sin x + 1 = 0$$
$$\sin x = \frac{1}{2} \text{ or } -1$$
$$x = \arcsin\left(\frac{1}{2}\right) \text{ or } \arcsin(-1)$$
$$\approx 0.52 \text{ or } -1.57$$

Of those solutions, only 0.52 is listed in the answer choices.

49. (C)—The entire surface area you're looking for consists of the lateral areas of the outside cylinder and the inside cylinder, plus the areas of the larger top and bottom circles, minus the areas of the smaller top and bottom circles. The lateral area of the outside cylinder is $2\pi rh = 2\pi(2)(6) = 24\pi$. The lateral area of the inside cylinder is $2\pi rh = 2\pi(1)(6) = 12\pi$. The areas of the larger top and bottom circles are each $\pi r^2 = \pi(2^2) = 4\pi$. And the areas of the smaller top and bottom circles are each $\pi r^2 = \pi(1^2) = \pi$. The total surface area, then, is:

$$24\pi + 12\pi + 2(4\pi) - 2(\pi) = 42\pi$$

50. (C)—Visualize the situation and/or make a few sketches. Try to imagine as many points of intersection as possible. Here's a way to get four:

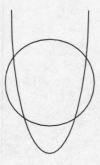

There's no way to get more.

Kaplan's Level IIC
Practice Test B

- The test that follows offers realistic practice for the SAT II: Mathematics Level IIC Test. To get the most out of it, you should take it under testlike conditions.

- Take the test in a quiet room with no distractions. Bring some No. 2 pencils and your calculator.

- Time yourself. You should spend no more than one hour on the 50 questions.

- Use the answer sheet to mark your answers.

- Answers and explanations follow the test.

- Scoring instructions are in "Compute your Level IIC Score" at the back of this section.

LEVEL IIC

KAPLAN PRACTICE TEST B
ANSWER SHEET

1 Ⓐ Ⓑ Ⓒ Ⓓ Ⓔ   14 Ⓐ Ⓑ Ⓒ Ⓓ Ⓔ   27 Ⓐ Ⓑ Ⓒ Ⓓ Ⓔ   40 Ⓐ Ⓑ Ⓒ Ⓓ Ⓔ
2 Ⓐ Ⓑ Ⓒ Ⓓ Ⓔ   15 Ⓐ Ⓑ Ⓒ Ⓓ Ⓔ   28 Ⓐ Ⓑ Ⓒ Ⓓ Ⓔ   41 Ⓐ Ⓑ Ⓒ Ⓓ Ⓔ
3 Ⓐ Ⓑ Ⓒ Ⓓ Ⓔ   16 Ⓐ Ⓑ Ⓒ Ⓓ Ⓔ   29 Ⓐ Ⓑ Ⓒ Ⓓ Ⓔ   42 Ⓐ Ⓑ Ⓒ Ⓓ Ⓔ
4 Ⓐ Ⓑ Ⓒ Ⓓ Ⓔ   17 Ⓐ Ⓑ Ⓒ Ⓓ Ⓔ   30 Ⓐ Ⓑ Ⓒ Ⓓ Ⓔ   43 Ⓐ Ⓑ Ⓒ Ⓓ Ⓔ
5 Ⓐ Ⓑ Ⓒ Ⓓ Ⓔ   18 Ⓐ Ⓑ Ⓒ Ⓓ Ⓔ   31 Ⓐ Ⓑ Ⓒ Ⓓ Ⓔ   44 Ⓐ Ⓑ Ⓒ Ⓓ Ⓔ
6 Ⓐ Ⓑ Ⓒ Ⓓ Ⓔ   19 Ⓐ Ⓑ Ⓒ Ⓓ Ⓔ   32 Ⓐ Ⓑ Ⓒ Ⓓ Ⓔ   45 Ⓐ Ⓑ Ⓒ Ⓓ Ⓔ
7 Ⓐ Ⓑ Ⓒ Ⓓ Ⓔ   20 Ⓐ Ⓑ Ⓒ Ⓓ Ⓔ   33 Ⓐ Ⓑ Ⓒ Ⓓ Ⓔ   46 Ⓐ Ⓑ Ⓒ Ⓓ Ⓔ
8 Ⓐ Ⓑ Ⓒ Ⓓ Ⓔ   21 Ⓐ Ⓑ Ⓒ Ⓓ Ⓔ   34 Ⓐ Ⓑ Ⓒ Ⓓ Ⓔ   47 Ⓐ Ⓑ Ⓒ Ⓓ Ⓔ
9 Ⓐ Ⓑ Ⓒ Ⓓ Ⓔ   22 Ⓐ Ⓑ Ⓒ Ⓓ Ⓔ   35 Ⓐ Ⓑ Ⓒ Ⓓ Ⓔ   48 Ⓐ Ⓑ Ⓒ Ⓓ Ⓔ
10 Ⓐ Ⓑ Ⓒ Ⓓ Ⓔ   23 Ⓐ Ⓑ Ⓒ Ⓓ Ⓔ   36 Ⓐ Ⓑ Ⓒ Ⓓ Ⓔ   49 Ⓐ Ⓑ Ⓒ Ⓓ Ⓔ
11 Ⓐ Ⓑ Ⓒ Ⓓ Ⓔ   24 Ⓐ Ⓑ Ⓒ Ⓓ Ⓔ   37 Ⓐ Ⓑ Ⓒ Ⓓ Ⓔ   50 Ⓐ Ⓑ Ⓒ Ⓓ Ⓔ
12 Ⓐ Ⓑ Ⓒ Ⓓ Ⓔ   25 Ⓐ Ⓑ Ⓒ Ⓓ Ⓔ   38 Ⓐ Ⓑ Ⓒ Ⓓ Ⓔ
13 Ⓐ Ⓑ Ⓒ Ⓓ Ⓔ   26 Ⓐ Ⓑ Ⓒ Ⓓ Ⓔ   39 Ⓐ Ⓑ Ⓒ Ⓓ Ⓔ

right

wrong

Use the answer key following the test to count up the number of questions you got right and the number you got wrong. (Remember not to count omitted questions as wrong.) "Compute Your Level IIC Score" at the back of this section will show you how to find your score.

LEVEL IIC

PRACTICE TEST B

50 Questions (1 hour)

Directions: For each question, choose the BEST answer from the choices given. If the precise answer is not among the choices, choose the one that best approximates the answer.

Notes:

(1) To answer some of these questions you will need a calculator. You must use at least a scientific calculator, but programmable and graphing calculators are also allowed.

(2) Figures in this test are drawn as accurately as possible UNLESS it is stated in a specific question that the figure is not drawn to scale. All figures are assumed to lie in a plane unless otherwise specified.

(3) The domain of any function f is assumed to be the set of all real numbers x for which $f(x)$ is a real number, unless otherwise indicated.

Reference Information: Use the following formulas as needed.

Right circular cone: If r = radius and h = height, then **Volume** $= \dfrac{1}{3}\pi r^2 h$; and if c = circumference of the base and ℓ = slant height, then **Lateral Area** $= \dfrac{1}{2}c\ell$.

Sphere: If r = radius, then **Volume** $= \dfrac{4}{3}\pi r^3$ and **Surface Area** $= 4\pi r^2$.

Pyramid: If B = area of the base and h = height, then **Volume** $= \dfrac{1}{3}Bh$.

---

1. If $\dfrac{x+y}{0.01} = 7$, then $\dfrac{1}{2x+2y} =$

    (A) 0.14   (B) 0.28   (C) 3.50   (D) 7.00   (E) 7.14

DO YOUR FIGURING HERE.

TURN TO THE NEXT PAGE.

LEVEL IIC

DO YOUR FIGURING HERE.

2. $\dfrac{\left(100^{12}\right)\left(10^4\right)}{10^2} =$

(A) 10^8 (B) 10^{14} (C) 10^{24} (D) 10^{26} (E) 10^{48}

3. If $\dfrac{x^2}{4} = \dfrac{6}{x}$, then $x =$

(A) 2.59   (B) 2.88   (C) 3.03   (D) 3.89   (E) 8.00

4. Which of the following is an equation of a line that will have points in all the quadrants except the first?

(A) $y = 2x$
(B) $y = 2x + 3$
(C) $y = 2x - 3$
(D) $y = -2x + 3$
(E) $y = -2x - 3$

5. If $b = 3 - a$ and $b \neq a$, then $\dfrac{a^2 - b^2}{b - a} =$

(A) 3   (B) 1   (C) 0   (D) –1   (E) –3

6. If $f(x) = ex + 2x$, then $f(\ln 2) =$

(A) 1.20
(B) 2.69
(C) 2.77
(D) 3.39
(E) 4.00

TURN TO THE NEXT PAGE.

LEVEL IIC

KAPLAN PRACTICE TEST B

7. In Figure 1, which of the following is the slope of line ℓ ?

   (A) -3

   (B) -2

   (C) $-\dfrac{1}{2}$

   (D) $\dfrac{1}{2}$

   (E) 2

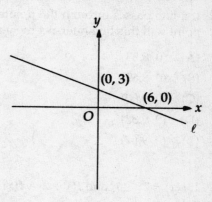

Figure 1

8. Which of the following is the complete solution set to the inequality $|x| + |x - 3| > 3$?

   (A) $\{x : x > 3 \text{ or } x < 0\}$

   (B) $\{x : -3 < x < 3\}$

   (C) $\{x : -3 > x\}$

   (D) $\{x : -3 < x\}$

   (E) $\{x : \text{The set of all real numbers}\}$

9. Which of the following is the solution set for $(3x - 6)(2 + x) < 0$?

   (A) $\{x: x < 2\}$

   (B) $\{x: x > 2\}$

   (C) $\{x: x > -2\}$

   (D) $\{x: x < -2 \text{ or } x > 2\}$

   (E) $\{x: -2 < x < 2\}$

TURN TO THE NEXT PAGE.

DO YOUR FIGURING HERE.

10. If a line passes through the points (5, 3) and (8, –1), at what point will this line intersect the y-axis?

    (A) (0, 8.33)

    (B) (0, 8.67)

    (C) (0, 9.00)

    (D) (0, 9.33)

    (E) (0, 9.67)

11. If $f(x) = 2x + 1$, and $f(x + 2) + f(x) = x$, what is the value of x ?

    (A) –2   (B) –1   (C) $-\dfrac{1}{2}$ (D) $\dfrac{1}{2}$ (E) 1

12. A certain stock begins the week trading at $87\dfrac{1}{2}$ per share.

    If the average gain for the next four days is $\dfrac{1}{2}$, by how much should the price of the stock increase during Friday so that the total gain for the stock during the entire five days is 5 percent?

    (A) $1\dfrac{3}{4}$ (B) $1\dfrac{7}{8}$ (C) $2\dfrac{1}{8}$ (D) $2\dfrac{1}{4}$ (E) $2\dfrac{3}{8}$

TURN TO THE NEXT PAGE.

KAPLAN PRACTICE TEST B

DO YOUR FIGURING HERE.

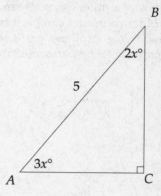

Figure 2

13. In Figure 2, what is the length of AC ?

    (A)   2.94

    (B)   3.49

    (C)   3.81

    (D)   4.05

    (E)   4.26

14. If $a = \sqrt[3]{t}$ and $b = t^2$, then $\dfrac{b}{a^5} =$

    (A) $t^{-\frac{1}{3}}$

    (B) $t^{\frac{1}{3}}$

    (C) $t^{\frac{5}{6}}$

    (D) $t^{\frac{6}{5}}$

    (E) $t^{\frac{10}{3}}$

15. If A, B, C, D, E, and F are 6 distinct points on the circumference of a circle, how many different chords can be drawn using any 2 of the 6 points?

    (A) 6   (B) 12   (C) 15   (D) 30   (E) 36

TURN TO THE NEXT PAGE.

DO YOUR FIGURING HERE.

16. A new computer can perform x calculations in y seconds and an older computer can perform r calculations in s minutes. If these two computers work simultaneously, how many calculations can be performed in t minutes?

(A) $t\left(\dfrac{x}{60y}+\dfrac{r}{s}\right)$

(B) $t\left(\dfrac{60x}{y}+\dfrac{r}{s}\right)$

(C) $t\left(\dfrac{x}{y}+\dfrac{r}{s}\right)$

(D) $t\left(\dfrac{x}{y}+\dfrac{60r}{s}\right)$

(E) $60t\left(\dfrac{x}{y}+\dfrac{r}{s}\right)$

17. Which of the following could be the equation of the parabola in Figure 3 ?

(A) $y = (x - 2)(x - 3)$
(B) $y = (x + 2)(x + 3)$
(C) $y = (x + 2)(x - 3)$
(D) $y = (x - 2)(x + 3)$
(E) $x = (y + 2)(y - 3)$

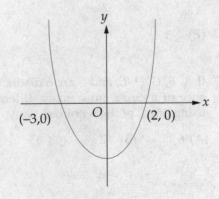

Figure 3

18. If $a + b = 15$, $b + c = 10$, and $a + c = 13$, which of the following is true?

(A) $a < b < c$
(B) $b < a < c$
(C) $c < b < a$
(D) $a < c < b$
(E) $b < c < a$

TURN TO THE NEXT PAGE.

DO YOUR FIGURING HERE.

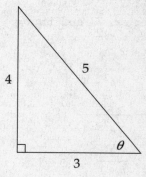

Figure 4

19. In Figure 4, $\dfrac{1}{\sin\theta} + \dfrac{1}{\cos\theta} =$

   (A) 0.75

   (B) 1.20

   (C) 1.43

   (D) 2.74

   (E) 2.92

20. Amanda goes to the toy store to buy 1 ball—either a football, basketball, or soccer ball—and 3 different board games. If the toy store is stocked with all types of balls and but only 6 different types of board games, how many different selections of 4 items can Amanda make consisting of 1 type of ball and 3 different board games?

   (A) 18   (B) 20   (C) 54   (D) 60   (E) 162

21. If point $P(3, 2)$ is rotated 90 degrees counterclockwise with respect to the origin, what will be its new coordinates?

   (A) (–2, 3)

   (B) (–2, –3)

   (C) (–3, 3)

   (D) (–3, 2)

   (E) (–3, –2)

TURN TO THE NEXT PAGE.

DO YOUR FIGURING HERE.

22. If $0 < x < \dfrac{\pi}{2}$, and $\tan x = \dfrac{a}{2}$, then $\cos x =$

   (A) $\dfrac{2}{\sqrt{a^2 - 4}}$

   (B) $\dfrac{a}{\sqrt{a^2 - 4}}$

   (C) $\dfrac{2}{a + 2}$

   (D) $\dfrac{2}{\sqrt{a^2 + 4}}$

   (E) $\dfrac{a}{\sqrt{a^2 + 4}}$

23. For what value of x will $f(x) = (1 - 2x)^2$ have the minimum value?

   (A) -1 (B) $-\dfrac{1}{2}$ (C) 0 (D) $\dfrac{1}{2}$ (E) 1

24. If a certain line intersects the origin and is perpendicular to the line with the equation $y = 2x + 5$ at point P, what is the distance from the origin to point P ?

   (A)   2.24

   (B)   2.45

   (C)   2.67

   (D)   3.89

   (E)   3.25

TURN TO THE NEXT PAGE.

DO YOUR FIGURING HERE.

25. If the volume of a cube is equal to the volume of a sphere, what is the ratio of the edge of the cube to the radius of the sphere?

    (A) 1.61

    (B) 2.05

    (C) 2.33

    (D) 2.45

    (E) 2.65

26. If $[x]$ represents the greatest integer less than or equal to x, what is the solution to the equation $1 - 2[x] = -3$?

    (A) $x = 2$

    (B) $2 \leq x < 3$

    (C) $2 < x \leq 3$

    (D) $2 < x < 3$

    (E) There is no solution.

27. Which of the following lists all and only the vertical asymptotes of the graph $y = \dfrac{x}{x^2 - 4}$?

    (A) $x = 2$ only

    (B) $y = 2$ only

    (C) $x = 2$ and $x = -2$

    (D) $y = 2$ and $y = -2$

    (E) $x = 2$, $x = -2$, and $x = 0$

28. If $\cos x \sin x = 0.22$, then $(\cos x - \sin x)^2 =$

    (A) 0

    (B) 0.11

    (C) 0.44

    (D) 0.56

    (E) 1.00

TURN TO THE NEXT PAGE.

29. If water is poured at a rate of 12 cubic meters per second into a half-empty rectangular tank with length 5 meters, width 3 meters, and height 25 meters, then how high, in meters, will the water level be after 9 seconds?

    (A)   6.0
    (B)   7.2
    (C)   18.5
    (D)   19.7
    (E)   The tank will be full and overflowing.

30. A circle centered at (3, 2) with radius 5 intersects the x-axis at which of the following x-coordinates?

    (A)   2.39
    (B)   4.58
    (C)   7.58
    (D)   8.00
    (E)   8.39

31. If $0 \leq x \leq \pi$, where is $\dfrac{\tan x}{\sin x}$ defined?

    (A) $0 \leq x \leq \pi$

    (B) $0 < x < \pi$

    (C) $0 < x < \dfrac{\pi}{2}$

    (D) $\dfrac{\pi}{2} \leq x \leq \pi$

    (E) $0 < x < \dfrac{\pi}{2}$ and $\dfrac{\pi}{2} < x < \pi$

TURN TO THE NEXT PAGE.

DO YOUR FIGURING HERE.

32. A rectangular box with an open top is constructed from cardboard to have a square base of area x^2 and height h. If the volume of this box is 50 cubic units, how many square units of cardboard, in terms of x, is needed to build this box?

(A) $5x^2$

(B) $6x^2$

(C) $\dfrac{200}{x} + x^2$

(D) $\dfrac{200}{x} + 2x^2$

(E) $\dfrac{250}{x} + 2x^2$

33. $\dfrac{(n + 2)! - (n + 1)!}{n!} =$

(A) $(n + 2)!$
(B) $(n + 1)!$
(C) $(n + 2)^2$
(D) $(n + 1)^2$
(E) n

TURN TO THE NEXT PAGE.

DO YOUR FIGURING HERE.

34. Bob wishes to borrow some money. He needs to defer to the following formula, where M is the monthly payment, r is the monthly decimal interest rate, P is the amount borrowed, and t is the number of months it will take to repay the loan:

$$M = \frac{rP}{1 - \left(\dfrac{1}{1+r}\right)^t}$$

If Bob secures a loan of $4,000 that he will pay back in 36 months with a monthly interest rate of 0.01, what is his monthly payment?

(A) $111.11

(B) $119.32

(C) $132.86

(D) $147.16

(E) $175.89

35. A particle is moving along the line $5y = -6x + 30$ at a rate of 2 units per second. If the particle starts at the y-intercept and moves to the right along this line, how many seconds will it take for the particle to reach the x-axis?

(A) 2.50

(B) 3.25

(C) 3.76

(D) 3.91

(E) 7.81

TURN TO THE NEXT PAGE.

DO YOUR FIGURING HERE.

36. In Figure 5, if the area of triangle *ABC* is 15, what is the length of *AC* ?

   (A)   2.1
   (B)   4.1
   (C)   6.2
   (D)   8.2
   (E)   9.6

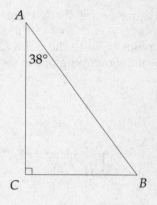

Figure 5

37. Which of the following functions has a range of $-1 < y < 1$?

   (A) $y = \sin x$

   (B) $y = \cos x$

   (C) $y = \dfrac{x}{1+x}$

   (D) $y = \dfrac{x^2}{1+x^2}$

   (E) $y = \dfrac{x}{\sqrt{1+x^2}}$

38. What is the sum of the infinite series $1 - \dfrac{1}{3} + \dfrac{1}{9} - \dfrac{1}{27} + \ldots$?

   (A) $\dfrac{2}{3}$ (B) $\dfrac{3}{4}$ (C) 1   (D) $\dfrac{4}{3}$ (E) $\dfrac{3}{2}$

TURN TO THE NEXT PAGE.

DO YOUR FIGURING HERE.

39. In Figure 6, the shaded region represents the set C of all points (x, y) such that $x^2 + y^2 \leq 1$. The transformation T maps the point (x, y) to the point $(2x, 4y)$ Which of the following shows the mapping of the set C by the transformation T ?

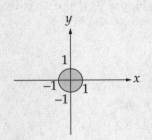

Figure 6

(A)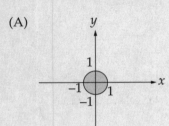

(B)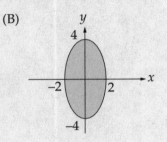

(C)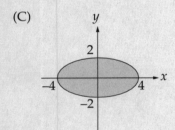

(D)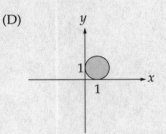

(E)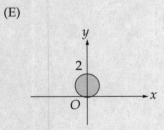

TURN TO THE NEXT PAGE.

DO YOUR FIGURING HERE.

40. $\lim\limits_{n \to \infty} \dfrac{1 - 2n^2}{5n^2 - n + 100} =$

   (A) -1

   (B) $-\dfrac{2}{5}$

   (C) $\dfrac{2}{5}$

   (D) 1

   (E)   No limit exists.

41. If $\log_2(x^2 - 3) = 5$, which of the following could be the value of x ?

   (A)   3.61
   (B)   4.70
   (C)   5.29
   (D)   5.75
   (E)   5.92

42. If 2 is a zero of the function $f(x) = 6x^3 - 11x^2 - 3x + 2$, what are the other zeroes?

   (A) $-\dfrac{1}{3}$ and $-\dfrac{1}{2}$

   (B) $-\dfrac{1}{3}$ and $\dfrac{1}{2}$

   (C) $\dfrac{1}{3}$ and $-\dfrac{1}{2}$

   (D) $\dfrac{1}{3}$ and $\dfrac{1}{2}$

   (E)   2 and 3

TURN TO THE NEXT PAGE.

43. In Figure 7, a circle of radius 1 is placed on an incline where point P, a point on the circle, has the coordinates $(-5, -5)$. The circle is rolled up the incline, and once the circle hits the origin, the circle is then rolled horizontally along the x-axis to the right. What is the x-coordinate of the point where P touches the incline or the x-axis for the fifth time (not including the starting point)?

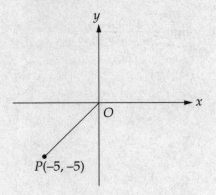

Figure 7

(A)   8.64

(B)   17.27

(C)   24.34

(D)   27.49

(E)   30.63

44. If $0 \leq x \leq 2\pi$ and $\sin x < 0$, which of the following must be true?

   I. $\cos x < 0$

  II. $\csc x < 0$

 III. $|\sin x + \cos x| > 0$

(A)   I only

(B)   II only

(C)   III only

(D)   I and II

(E)   II and III

45. If $i^2 = -1$, which of the following is a square root of $8 - 6i$?

(A) $3 - i$

(B) $3 + i$

(C) $3 - 4i$

(D) $4 - 3i$

(E) $4 + 3i$

TURN TO THE NEXT PAGE.

KAPLAN PRACTICE TEST B

DO YOUR FIGURING HERE.

46. Figure 8 shows rectangle $ABCD$. Points A and D are on the parabola $y = 2x^2 - 8$, and points B and C are on the parabola $y = 9 - x^2$. If point B has coordinates $(-1.50, 6.75)$, what is the area of rectangle $ABCD$?

(A) 12.50

(B) 17.50

(C) 22.75

(D) 26.50

(E) 30.75

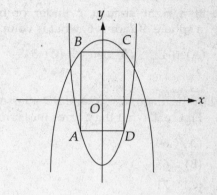

Figure 8

47. If $x \geq 0$ and $\arcsin x = \arccos (2x)$, then $x =$

(A) 0.866

(B) 0.707

(C) 0.500

(D) 0.447

(E) 0.245

48. If $f(x) = \dfrac{1}{2}x - 4$ and $f(g(x)) = g(f(x))$, which of the following can be $g(x)$?

I. $2x - \dfrac{1}{4}$

II. $2x + 8$

III. $\dfrac{1}{2}x - 4$

(A) I only

(B) II only

(C) III only

(D) II and III only

(E) I, II, and III

TURN TO THE NEXT PAGE.

LEVEL IIC

KAPLAN PRACTICE TEST B

49. If a right circular cylinder of height 10 is inscribed in a sphere of radius 6, what is volume of the cylinder?

    (A) 104   (B) 346   (C) 545   (D) 785   (E) 1,131

50. If the diagonals *AC* and *BD* intersect at point *P* in the cube in Figure 9, what the degree measure of angle *APB* ?

    (A) 60
    (B) 65
    (C) 71
    (D) 83
    (E) 90

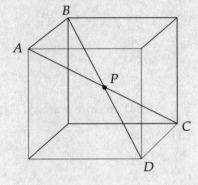

Figure 9

STOP! END OF TEST. DO NOT TURN THE PAGE UNTIL YOU ARE READY TO CHECK YOUR ANSWERS.

Turn the page for
Level IIC Test B
answers and explanations.

Level IIC Test B—Answer Key

1. E	11. A	21. A	31. E	41. E
2. D	12. E	22. D	32. C	42. C
3. B	13. A	23. D	33. D	43. C
4. E	14. B	24. A	34. C	44. B
5. E	15. C	25. A	35. D	45. A
6. D	16. B	26. B	36. C	46. E
7. C	17. D	27. C	37. E	47. D
8. A	18. C	28. D	38. B	48. D
9. E	19. E	29. D	39. B	49. B
10. E	20. D	30. C	40. B	50. C

1. **(E)**—If $\dfrac{x+y}{0.01} = 7$, then $x + y = 7(0.01) = 0.07$.
Therefore:

$$\frac{1}{2x+2y} = \frac{1}{2(x+y)}$$

$$= \frac{1}{2(0.07)} = \frac{1}{0.14} \approx 7.14$$

2. **(D)**—Put the whole thing in terms of a power of 10. $(100)^{12} = (10^2)^{12} = 10^{24}$. Therefore:

$$\frac{(100)^{12}(10)^4}{(10)^2} = \frac{\left(10^{24}\right)\left(10^4\right)}{10^2}$$

$$= \frac{10^{28}}{10^2} = 10^{28-2} = 10^{26}$$

3. **(B)**—Cross multiply:

$$\frac{x^2}{4} = \frac{6}{x}$$

$$x^2 \times x = 4 \times 6$$

$$x^3 = 24$$

Now use your calculator to find the cube root of 24:

$$x = \sqrt[3]{24} \approx 2.88$$

4. **(E)**—The answer choices are all linear equations in convenient $y = mx + b$ form. A line that has points in all quadrants but the first is a line that crosses the y-axis below the origin and heads downhill from there—in other words, a line with both a negative y-intercept and a negative slope. (C) and (E) have negative y-intercepts. (D) and (E) have negative slopes. Only (E) has both.

5. **(E)**—Factor the numerator and look for something you can cancel with the denominator:

$$\frac{a^2 - b^2}{b-a} = \frac{(a-b)(a+b)}{b-a}$$

$$= \frac{-(b-a)(a+b)}{b-a}$$

$$= -(a+b)$$

It's given that $b = 3 - a$, which is just another way of saying $a + b = 3$, so:

$$-(a+b) = -3$$

6. (D)—To find the value of $f(x)$ for a particular value of x, plug it into the definition:

$$f(x) = e^x + 2x$$
$$f(\ln 2) = e^{\ln 2} + 2\ln 2$$

You'll need your calculator to evaluate part of this expression, but you should realize, without a calculator, that $e^{\ln 2}$ is 2. Use your calculator to find that $2\ln 2 \approx 1.39$, and therefore:

$$f(\ln 2) = e^{\ln 2} + 2\ln 2$$
$$\approx 2 + 1.39 = 3.39$$

7. (C)—You can use the two given points to figure out the slope:

$$\text{Slope} = \frac{y_2 - y_1}{x_2 - x_1}$$
$$= \frac{3 - 0}{0 - 6} = \frac{3}{-6} = -\frac{1}{2}$$

8. (A)—Think about the three different cases.

Case 1, when $x \geq 3$:

$$|x| + |x - 3| > 3$$
$$x + x - 3 > 3$$
$$2x - 3 > 3$$
$$2x > 6$$
$$x > 3$$

So all numbers greater than 3 satisfy the inequality, but $x = 3$ itself does not.

Case 2, when $0 \leq x < 3$:

$$|x| + |x - 3| > 3$$
$$x + (-x) + 3 > 3$$
$$3 > 3$$

So nothing between 0 and 3 satisfies the inequality.

Case 3, when $x < 0$:

$$|x| + |x - 3| > 3$$
$$(-x) + (-x) + 3 > 3$$
$$-2x + 3 > 3$$
$$-2x > 0$$
$$x < 0$$

So all negative numbers satisfy the inequality, and the complete solution set is $\{x : x > 3 \text{ or } x < 0\}$.

9. (E)—If the product of $(3x - 6)$ and $(2 + x)$ is negative, then one of the two factors is negative and the other is positive. There are two cases:

Case 1, when $3x - 6 < 0$ and $2 + x > 0$:

$$3x - 6 < 0$$
$$3x < 6$$
$$x < 2$$

and:

$$2 + x > 0$$
$$x > -2$$

So, all x such that $-2 < x < 2$ satisfy the inequality.

Case 2, when $3x - 6 > 0$ and $2 + x < 0$:

$$3x - 6 > 0$$
$$3x > 6$$
$$x > 2$$

and:

$$2 + x < 0$$
$$x < -2$$

There are no numbers that are both less than –2 and greater than 2, so the numbers that work in Case 1 are the complete solution set.

10. (E)—You can use the two given points to find the slope:

$$\text{Slope} = \frac{y_2 - y_1}{x_2 - x_1}$$
$$= \frac{-1 - 3}{8 - 5} = \frac{-4}{3} = -\frac{4}{3}$$

Next plug the point $(0, y)$ and either one of the given points into the same formula:

$$\text{Slope} = \frac{y_2 - y_1}{x_2 - x_1}$$

$$-\frac{4}{3} = \frac{3 - y}{5 - 0}$$

$$-\frac{4}{3} = \frac{3 - y}{5}$$

$$-3(3 - y) = (4)(5)$$

$$3 - y = \frac{20}{-3}$$

$$-y = \frac{20}{-3} - 3$$

$$y = \frac{20}{3} + 3 \approx 9.67$$

11. (A)—Plug both x and $x + 2$ into the definition:

$$f(x) = 2x + 1$$
$$f(x + 2) = 2(x + 2) + 1$$

Then set the sum equal to x and solve:

$$f(x + 2) + f(x) = x$$
$$2(x + 2) + 1 + 2x + 1 = x$$
$$2x + 4 + 1 + 2x + 1 - x = 0$$
$$3x + 6 = 0$$
$$3x = -6$$
$$x = -2$$

12. (E)—First find 5 percent of $87\frac{1}{2}$:

$$(.05)(87.5) = 4.375$$

An average daily gain of $\frac{1}{2}$ for four days means a net gain of 2. To go up 5 percent for the week, the stock needs to go up another 4.375 – 2 = 2.375, which is the same as $2\frac{3}{8}$.

13. (A)—It's a right triangle, so the two acute angles add up to 90 degrees. Since their degree measures are in a 2-to-3 ratio, they must be $\frac{2}{5}$ of 90 and $\frac{3}{5}$ of 90, or 36 and 54:

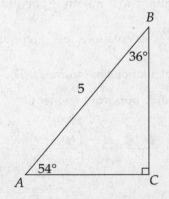

The side AC you're looking for is opposite the 36° angle, so:

$$\frac{AC}{5} = \sin 36°$$

Use your calculator to find that $\sin 36° \approx 0.588$. Therefore:

$$\frac{AC}{5} \approx 0.588$$
$$AC \approx 5(0.588) = 2.94$$

14. **(B)**—Plug $a = \sqrt[3]{t} = t^{\frac{1}{3}}$ and $b = t^2$ into the expression:

$$\frac{b}{a^5} = \frac{t^2}{\left(t^{\frac{1}{3}}\right)^5} = \frac{t^2}{t^{\frac{5}{3}}} = t^{2-\frac{5}{3}} = t^{\frac{1}{3}}$$

15. **(C)**—Each point is connected to five other points to make 5 chords per point. But this counts every chord twice—AB is indistinguishable from BA—so after you multiply 6 by 5, you have to divide by 2, yielding 15.

16. **(B)**—Take the computers one at a time. Put everything in terms of seconds: t minutes is $60t$ seconds and s minutes is $60s$ seconds. The new computer performs $\frac{x}{y}$ calculations per second, which is $\frac{60tx}{y}$ calculations in t minutes. The old computer performs $\frac{r}{60s}$ calculations per second, which is $\frac{60tr}{60s} = \frac{tr}{s}$ calculations in t minutes. The number of calculations they perform together is:

$$\frac{60tx}{y} + \frac{tr}{s} = t\left(\frac{60x}{y} + \frac{r}{s}\right)$$

17. **(D)**—Try the two given points in the choices. The point $(-3, 0)$ satisfies (B) and (D). The point $(2, 0)$ satisfies (A) and (D). The only choice that works both times is (D).

18. **(C)**—You want to rank a, b, and c. Actually, because you're given three equations, you can solve for the three unknowns. Subtract the equation $b + c = 10$ from the equation $a + b = 15$ and you get: $a - c = 5$. Now add that to the equation $a + c = 13$ and you get: $2a = 18$, or $a = 9$. Now you can plug that value of a back into the appropriate equations to find that $b = 6$ and $c = 4$. So the correct ranking is $c < b < a$.

19. **(E)**—Sine is opposite over hypotenuse and cosine is adjacent over hypotenuse, so $\sin\theta = \frac{4}{5}$ and $\cos\theta = \frac{3}{5}$. Therefore:

$$\frac{1}{\sin\theta} + \frac{1}{\cos\theta} = \frac{5}{4} + \frac{5}{3} \approx 1.25 + 1.67 = 2.92$$

20. **(D)**—Amanda has 3 choices for the ball. As for the board games, she wants to choose 3 out of 6, so the number of board game combinations is:

$$_6C_3 = \frac{6!}{(6-3)!3!} = \frac{6 \cdot 5 \cdot 4 \cdot 3 \cdot 2 \cdot 1}{3 \cdot 2 \cdot 1 \cdot 3 \cdot 2 \cdot 1} = 20$$

For each of 20 board game combinations there are three ball choices, so the total number of options Amanda faces is $3 \times 20 = 60$.

21. **(A)**—Sketch a diagram:

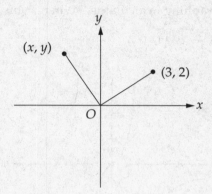

Add a couple of perpendiculars and you'll make a couple of congruent triangles:

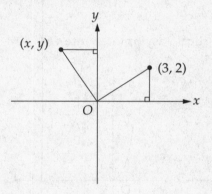

To get to the original point $P(3, 2)$, you go right 3 and up 2. To get to the new point (x, y) you go left 2 and up 3, so its coordinates are $(-2, 3)$.

22. **(D)**—Since x is an acute angle, use a right triangle to solve this one. Since $\tan x = \dfrac{a}{2}$, and tangent is opposite over adjacent, you could label the opposite side a and the adjacent side 2:

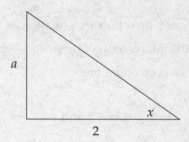

Then, by the Pythagorean theorem, the hypotenuse is $\sqrt{a^2 + 4}$: $\square$

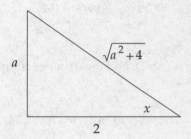

Cosine is adjacent over hypotenuse, so $\cos x$ is $\dfrac{2}{\sqrt{a^2 + 4}}$. $\square$

23. **(D)**—The expression $(1 - 2x)^2$ is something squared. A real number squared cannot be smaller than 0. This function will have its minimum value when the part inside the parentheses is zero:

$$1 - 2x = 0$$
$$-2x = -1$$
$$x = \frac{1}{2}$$

24. **(A)**—If the two lines are perpendicular, then their slopes are negative reciprocals. The slope of $y = 2x + 5$ is 2, so the slope of the other line is $-\dfrac{1}{2}$. That line goes through the origin, so its y-inter-

cept is 0, and therefore its equation is $y = -\dfrac{1}{2}x$. Point P—the intersection of these lines—is the point that satisfies both equations:

$$y = 2x + 5 \text{ and } y = -\frac{1}{2}x$$

$$2x + 5 = -\frac{1}{2}x$$
$$4x + 10 = -x$$
$$5x = -10$$
$$x = -2$$
$$y = -\frac{1}{2}x = -\frac{1}{2}(-2) = 1$$

So point P is $(-2, 1)$, and the distance from the origin to point P is:

$$OP = \sqrt{(-2)^2 + 1^2} = \sqrt{5} \approx 2.24$$

25. (A)—Set the formulas equal and solve for $\dfrac{e}{r}$:

$$e^3 = \frac{4}{3}\pi r^3$$

$$\frac{e^3}{r^3} = \frac{4}{3}\pi$$

$$\frac{e}{r} = \sqrt[3]{\frac{4}{3}\pi} \approx 1.61$$

26. (B)—First solve for $[x]$:

$$1 - 2[x] = -3$$
$$-2[x] = -4$$
$$[x] = 2$$

$[x]$ will equal 2 whenever $2 \le x < 3$.

27. (C)—The easiest way to do this one is to use your graphing calculator. When you graph $y = \dfrac{x}{x^2 - 4}$, you get:

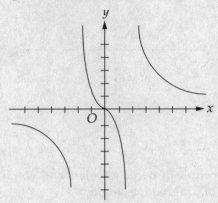

The vertical asymptotes are at $x = -2$ and $x = 2$:

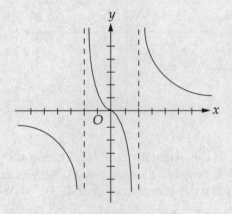

28. (D)—Expand the expression you're solving for and see what happens:

$$(\cos x - \sin x)^2 = \cos^2 x - 2\cos x \sin x + \sin^2 x$$
$$= \left(\cos^2 x + \sin^2 x\right) - 2\cos x \sin x$$
$$= 1 - 2(0.22) = 1 - 0.44 = 0.56$$

29. (D)—Sketch (or visualize) the situation. Before more water is added, the water level is half of 25, or 12.5 feet:

As water is poured in, the rectangular solid that represents the water gets taller, while the dimensions of the base stay the same. As water is poured in for 9 seconds at the rate of 12 cubic meters per second, the amount added is $9 \times 12 = 108$ cubic meters. With a base of 3 meters by 5 meters and the 108 cubic meters of added volume, you can figure out what the added depth is:

$$\text{Added volume} = \text{length} \times \text{width} \times \text{added depth}$$

$$108 = 5 \times 3 \times \text{added depth}$$

$$\text{added depth} = \frac{108}{15} = 7.2$$

That's on top of the pre-existing 12.5 meters. The new water level is $12.5 + 7.2 = 19.7$ meters.

30. (C)—The equation for a circle centered at (3, 2) and with radius $r = 5$ is:

$$(x-3)^2 + (y-2)^2 = 25$$

To find where the graph of this equation intersects the x-axis, set $y = 0$ and solve for x:

$$(x-3)^2 + (0-2)^2 = 25$$

$$(x-3)^2 + 4 = 25$$

$$(x-3)^2 = 21$$

$$x - 3 = \pm\sqrt{21}$$

$$x = 3 \pm \sqrt{21} \approx -1.58 \text{ or } 7.58$$

31. (E)—Between 0 and π, tangent is defined for anything but $\frac{\pi}{2}$. Sine is defined for all values of x, but since the sine is in the denominator here, the expression $\frac{\tan x}{\sin x}$ will be undefined when $\sin x = 0$, which is when $x = 0$ or π. So the expression $\frac{\tan x}{\sin x}$ is defined when $0 < x < \frac{\pi}{2}$ or $\frac{\pi}{2} < x < \pi$.

32. (C)—Use the given volume = 50 to get h in terms of x:

$$\text{Volume} = \text{length} \times \text{width} \times \text{height}$$

$$50 = x \cdot x \cdot h$$

$$h = \frac{50}{x^2}$$

The area you're looking for is equal to four lateral faces and one bottom face. Each lateral face has area $xh = x\dfrac{50}{x^2} = \dfrac{50}{x}$, and the base has area x^2, so the total area you're looking for is:

$$4\left(\frac{50}{x}\right) + x^2 = \frac{200}{x} + x^2$$

33. **(D)**—Expand, cancel, and simplify:

$$\frac{(n+2)!-(n+1)!}{n!} = \frac{(n+2)!}{n!} - \frac{(n+1)!}{n!}$$
$$= (n+2)(n+1) - (n+1)$$
$$= n^2 + 3n + 2 - n - 1$$
$$= n^2 + 2n + 1$$

That's the same as (D) $(n+1)^2$.

34. **(C)**—Plug $P = 4,000$, $t = 36$, and $r = 0.01$ into the formula and calculate:

$$M = \frac{rP}{1 - \left(\frac{1}{1+r}\right)^t} = \frac{(0.01)(4,000)}{1 - \left(\frac{1}{1+0.01}\right)^{36}} \approx 132.86$$

35. **(D)**—First find the distance from the starting point to the endpoint. The starting point is the y-intercept, which is easy to get if you put the given equation in slope-intercept form:

$$5y = -6x + 30$$
$$y = -\frac{6}{5}x + 6$$

The y-intercept is the point $(0, 6)$. The x-intercept is the point at which $y = 0$:

$$5(0) = -6x + 30$$
$$6x = 30$$
$$x = 5$$

So, the x-intercept is the point $(5, 0)$. The distance between those points is:

$$\text{Distance} = \sqrt{(y_2 - y_1)^2 + (x_2 - x_1)^2}$$
$$= \sqrt{(0-6)^2 + (5-0)^2}$$
$$= \sqrt{36 + 25}$$
$$= \sqrt{61} \approx 7.81$$

At a rate of 2 units per second, it will take $\frac{7.81}{2}$ seconds to travel 7.81 units. That's about 3.91.

36. **(C)**—Use the given angle to find the ratio of the legs:

$$\frac{BC}{AC} = \tan 38° \approx 0.781$$

If you call AC (the leg you're looking for) x, then you can call leg BC $0.781x$. Half the product of the legs is the area, or 15:

$$\text{Area} = \frac{1}{2}(\text{leg}_1)(\text{leg}_2)$$
$$15 = \frac{1}{2}(x)(0.781x)$$
$$0.781x^2 = 30$$
$$x^2 = \frac{30}{0.781} \approx 38.4$$
$$x \approx 6.2$$

37. **(E)**—You should know what the ranges are of

(A) $y = \sin x$ and (B) $y = \cos x$. They're both $-1 \le y \le 1$. They both include -1 and 1 themselves, so you can eliminate (A) and (B). The upper limit of (C) is 1, but (C) can be much smaller than -1, such as when $x = -0.99$. So you can eliminate (C). (D) can never be less than zero because the numerator and denominator are nonnegative for all possible x. The answer is (E) because as x gets very large, $\dfrac{x}{\sqrt{1+x^2}}$ approaches $\dfrac{x}{\sqrt{x^2}}$ or $\dfrac{x}{x} = 1$, so the value of the whole expression approaches 1; and as x gets very small (that is, very negative), $\dfrac{x}{\sqrt{1+x^2}}$ approaches $\dfrac{x}{\sqrt{x^2}} = \dfrac{x}{|x|} = -1$, so the value of the whole expression approaches -1.

38. **(B)**—Use the formula for the sum of an infinite geometric series. Here the first term a is 1 and the ratio r is $-\dfrac{1}{3}$:

$$S = \frac{a}{1-r} = \frac{1}{1-\left(-\dfrac{1}{3}\right)} = \frac{1}{\dfrac{4}{3}} = \frac{3}{4}$$

39. **(B)**—Think about what happens to the indicated intercepts when all x's are doubled and all y's are quadrupled. The x-intercepts become 2 and -2, and the y-intercepts become 4 and -4. Choice (B) fits.

40. **(B)**—To find the limit of this expression as n approaches infinity, think about what happens as n gets extremely large. What happens is that the n^2 terms become so huge that they dwarf all other terms into insignificance. So you can think of the expression as, in effect, $\dfrac{-2n^2}{5n^2} = -\dfrac{2}{5}$.

41. **(E)**—If $\log_2(x^2 - 3) = 5$, then $x^2 - 3 = 2^5 = 32$:

$$x^2 - 3 = 32$$
$$x^2 = 35$$
$$x = \pm\sqrt{35} \approx \pm 5.92$$

42. **(C)**—If 2 is a zero, then $x - 2$ is a factor. Factor that out of the polynomial $6x^3 - 11x^2 - 3x + 2$:

$$\frac{6x^3 - 11x^2 - 3x + 2}{x - 2} = 6x^2 + x - 1$$

Now you have a quadratic equation:

$$6x^2 + x - 1 = 0$$
$$(3x - 1)(2x + 1) = 0$$
$$3x - 1 = 0 \text{ or } 2x + 1 = 0$$
$$x = \frac{1}{3} \text{ or } -\frac{1}{2}$$

43. **(C)**—With 5 rotations, the circle covers a total distance equal to 5 times the circumference. The circumference of a circle of radius 1 is 2π, so 5 times that is 10π. Of that total distance 10π, $5\sqrt{2}$ is on the incline. The other $10\pi - 5\sqrt{2}$ is the distance traveled along the x-axis, so the x-coordinate of the endpoint is $10\pi - 5\sqrt{2} \approx 24.34$.

44. **(B)**—Sine is negative in the third and fourth quadrants. Cosine is negative in the second and third, so I is not necessarily true. Cosecant is 1 over sine, so where sine is negative, so is cosecant. Therefore II must be true. As for III: Absolute values are always nonnegative. The issue here is whether

$\sin x + \cos x$ can equal 0. In fact, if x is $\dfrac{7\pi}{4}$, then $\sin x = -\dfrac{\sqrt{2}}{2}$ and $\cos x = \dfrac{\sqrt{2}}{2}$; so $\sin x + \cos x = 0$ when $x = \dfrac{7\pi}{4}$.

45. (A)—Square the answer choices until you find one that gives you $8 - 6i$. Start with choice (A):

$$(3-i)^2 = (3-i)(3-i)$$
$$= 9 - 6i + i^2$$
$$= 9 - 6i - 1$$
$$= 8 - 6i$$

That's it.

46. (E)—To find the area of the rectangle, you need the base and height. Both parabolas are symmetric with respect to the y-axis, so if the x-coordinate of point B is –1.50, the length of segment BC, which is the base of the rectangle, is twice 1.50, or 3. To find the height of the rectangle, use the y-coordinate of point B and find the y-coordinate of point A. That's the value of y when you plug $x = -1.50$ into the equation $y = 2x^2 - 8$:

$$y = 2(-1.50)^2 - 8 = -3.5$$

So the height of the rectangle is 6.75 + 3.5 = 10.25, and the area of the rectangle is $3 \times 10.25 = 30.75$.

47. (D)—Take the sine of both sides:

$$\arcsin x = \arccos (2x)$$
$$\sin(\arcsin x) = \sin[\arccos(2x)]$$
$$x = \sin[\arccos(2x)]$$

The cosine and sine of an angle are related by the Pythagorean identity $\sin^2 x + \cos^2 x = 1$, therefore, for an angle whose cosine is $2x$, the sine is $\sqrt{1-(2x)^2} = \sqrt{1-4x^2}$:

$$x = \sin[\arccos(2x)]$$
$$x = \sqrt{1-4x^2}$$
$$x^2 = 1 - 4x^2$$
$$5x^2 = 1$$
$$x^2 = \frac{1}{5}$$
$$x = \sqrt{\frac{1}{5}} \approx 0.447$$

48. (D)—If $f(g(x)) = g(f(x))$, then it doesn't matter which function you apply first, you should get the same result. This will happen under two sets of circumstances. The first is when the functions are inverses, in which case both $f(g(x))$ and $g(f(x))$ will get you back to the x you started with. The second is when the two functions are identical, in which case $f(g(x))$ and $g(f(x))$ are identical. II is the inverse and III is the identical function, so II and III are possible.

49. **(B)**—Sketch or visualize the situation:

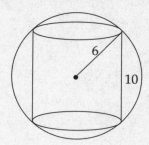

To get the volume of this cylinder, you need, in addition to the given height, the radius of the base. You can construct a right triangle and use the Pythagorean theorem to find that radius:

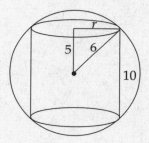

$$r = \sqrt{6^2 - 5^2} = \sqrt{11}$$

Now plug $r = \sqrt{11}$ and $h = 10$ into the cylinder volume formula:

$$\text{Volume} = \pi r^2 h = \pi\left(\sqrt{11}\right)^2 (10)$$
$$= 110\pi \approx 346$$

50. **(C)**—Look at rectangle $ABCD$:

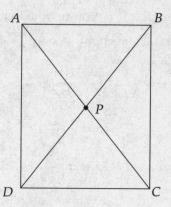

Say the edges of the cube are 1. Then the height of rectangle $ABCD$ is $\sqrt{2}$:

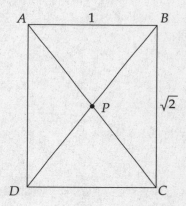

Divide triangle ABP into right triangles:

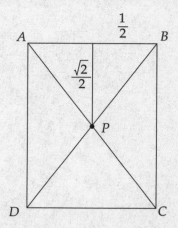

Half the angle APB you're looking for is the angle that has a tangent of $\frac{1}{2}$ over $\frac{\sqrt{2}}{2}$:

$$\tan\left(\frac{x}{2}\right) = \frac{\frac{1}{2}}{\frac{\sqrt{2}}{2}} = \frac{1}{\sqrt{2}}$$

$$\frac{x}{2} = \arctan\left(\frac{1}{\sqrt{2}}\right)$$

$$x = 2\arctan\left(\frac{1}{\sqrt{2}}\right) \approx 71°$$

COMPUTE YOUR LEVEL IIC SCORE

Step 1: Figure out your raw score. Refer to your answer sheet for the number right and the number wrong on the practice test you're scoring. (If you haven't checked your answers, do that now, using the answer key that follows the test.) You can use the chart below to figure out your raw score. Multiply the number wrong by .25 and subtract the result from the number right. Round the result to the nearest whole number. This is your raw score.

LEVEL IIC TEST A

NUMBER RIGHT	NUMBER WRONG	RAW SCORE
☐	– (.25 x ☐)	= ☐ (ROUNDED)

LEVEL IIC TEST B

NUMBER RIGHT	NUMBER WRONG	RAW SCORE
☐	– (.25 x ☐)	= ☐ (ROUNDED)

Step 2: Find your practice test score. Find your raw score in the left column of the table below. The score in the right column is your Level IIC Practice Test score.

Find Your Practice Test Score

Raw	Scaled	Raw	Scaled	Raw	Scaled	Raw	Scaled	Raw	Scaled	Raw	Scaled
50	800	39	740	28	630	17	540	6	420	–5	280
49	800	38	730	27	620	16	530	5	410	–6	260
48	800	37	720	26	610	15	520	4	400	–7	250
47	800	37	710	25	600	14	520	3	380	–8	240
46	800	35	700	24	590	13	510	2	370	–9	220
45	800	34	690	23	580	12	500	1	350	–10	210
44	790	33	680	22	580	11	490	0	340	–11	200
43	780	32	670	21	570	10	480	–1	330	–12	200
42	770	31	660	20	560	9	460	–2	310		
41	760	30	650	19	550	8	450	–3	300		
40	750	29	640	18	550	7	440	–4	290		

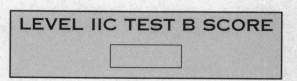

LEVEL IIC TEST A SCORE

LEVEL IIC TEST B SCORE

A note on your practice test scores: Don't take these scores too literally. Practice test conditions cannot precisely mirror real test conditions. Your actual SAT II: Mathematics Subject Test score will almost certainly vary from your practice test scores. Your scores on the practice tests will give you a rough idea of your range on the actual exam.

Robert Stanton has degrees in Slavic Languages and Literatures from Harvard and Yale. While at Yale, he taught his first SAT preparation course, and since then he has helped thousands of high school and college students do their best on such tests as the SAT, ACT, GRE, GMAT, and LSAT. Bob actually enjoys taking standardized tests, and he has made a career out of sharing that expertise and enthusiasm with his students. Bob is also a muralist, gourmet cook, world traveler, and inveterate first-nighter at the Metropolitan Opera.

ABOUT KAPLAN

COME TO US FOR THE BEST PREP

EDUCATIONAL CENTERS

"How can you help me?"

There are points in life when you need to reach an important goal. Whether you want a high score on a critical test, admission to a competitive college, funding for school, or career success, Kaplan is the best source to help get you there. One of the nation's premier educational companies, Kaplan has already helped millions of students get ahead through our legendary courses and expanding catalog of products and services.

"I have to ace this test!"

The world leader in test preparation, Kaplan will help you get a higher score on standardized tests such as the PSAT, SAT, and ACT for college, the LSAT, MCAT, GMAT, and GRE for graduate school, professional licensing exams for medicine, nursing, dentistry, and accounting, and specialized exams for international students and professionals.

Kaplan's courses are recognized worldwide for their high-quality instruction, state-of-the-art study tools and up-to-date, comprehensive information. With more than 160 permanent centers and 1,000 satellite classrooms, Kaplan enrolls more than 150,000 students annually in its live courses.

"How can I pay my way?"

As the price of higher education continues to skyrocket, it's vital to get your share of financial aid and figure out how you're going to pay for school. Kaplan's financial aid resources simplify the often bewildering application process and show you how you can afford to attend the college or graduate school of your choice.

Through KapLoan, The Kaplan Student Loan Information Program, we can help students get key information and advice about educational loans for college and graduate school. Through an affiliation with one of the nation's largest student loan providers, you can access valuable information and guidance on federally insured parent and student loans. Kaplan directs you to the financing you need to reach your educational goals.

"Can you help me find a good school?"

Kaplan offers expert advice on selecting a college, graduate school, or professional school. We can also show you how to maximize your chances of acceptance at the school of your choice.

"But then I have to get a great job!"

Whether you're a student or a grad, we can help you find a job that matches your interests. Kaplan can assist you by providing helpful assessment tests, job and employment data, recruiting services, and expert advice on how to land the right job. Our subsidiary, Crimson & Brown Associates, is the leading collegiate recruiting firm helping top-tier companies attract hard-to-find candidates.

Kaplan has the tools!

For students headed to college and graduate school, Kaplan offers the best-written, easiest-to-use **books.** Our growing library of titles includes guides for test preparation, selection, admissions, and financial aid.

Kaplan sets the standard for educational **software** with innovative products for college and graduate school preparation. Our bestselling RoadTrip™ test-prep software offers personalized study programs with practice tests and performance analysis. Our software library also offers the ultimate multimedia college guide, financial aid search resources, and tools for accelerating your academic and career success.

Helpful **videos** demystify college admissions and the SAT by leading the viewer on entertaining and irreverent road trips across America. Hitch a ride with Kaplan's *Secrets to College Admission* and *Secrets to SAT Success.*

Kaplan offers a variety of services **online** through sites on the Internet, America Online, and The MSN™ online service. Students can access information on testing, admissions, and careers; fun contests and special promotions; live events; bulletin boards; links to helpful sites; and plenty of downloadable files, games, and software. Kaplan Online is the ultimate student resource.

KAPLAN

KAPLAN

Want more information about our services, products, or the nearest Kaplan educational center?

--- **HERE** ---

Call our nationwide toll-free numbers:

1–800–KAP–TEST
(for information on our live courses)

1–800–KAP–ITEM
(for information on our products)

1–888–KAP–LOAN*
(for information on student loans)

Connect with us in cyberspace:
On **AOL**, keyword **"Kaplan"**
On the Internet's World Wide Web, open **"http://www.kaplan.com"**
On The **MSN**™** online service, Go Word **"KAPLAN"**
Via E-mail, **"info@kaplan.com"**

Write to:
Kaplan Educational Centers
888 Seventh Avenue
New York, NY 10106